The Use of Matrix Reference Materials
in Environmental Analytical Processes

The Use of Matrix Reference Materials in Environmental Analytical Processes

Edited by

A. Fajgelj

International Atomic Energy Agency, Seibersdorf, Austria

M. Parkany

Digart Ltd, Geneva, Switzerland

ROYAL SOCIETY OF CHEMISTRY

Proceedings of the Workshop on 'Proper Use of Environmental Matrix Reference Materials' held in Berlin on 22–23 April 1999

Chem
QD
131
.V84
1999

Special Publication No. 238

ISBN 0-85404-739-5

A catalogue record for this book is available from the British Library

Published by The Royal Society of Chemistry,
Thomas Graham House, Science Park, Milton Road,
Cambridge CB4 0WF, UK

For further information see our web site at www.rsc.org

Printed by MPG Books Ltd, Bodmin, Cornwall, UK

Foreword

Dr. John W. Jost

Executive Director, IUPAC

This is the fourth book in a series based on the work of the IUPAC Working Party on Harmonization of Quality Assurance Schemes for Analytical Laboratories. The papers contained in this volume were presented at a workshop on the Proper Use of Environmental Matrix Reference Materials, 22-23 April 1999, Berlin, Germany. The workshop was organized in conjunction with the ISO Committee on Reference Materials, the Bundesanstalt für Materialforschung und -prüfung and Eurolab-Deutschland. This activity is an excellent example of the kind of activity envisaged by the founders of IUPAC in 1919.

The founders of IUPAC were concerned that variations in terminology, symbols and test methods among different countries, and even among different industries, were creating barriers to trade and the growth of the chemical industry. The purpose of IUPAC from its inception has been to promote voluntary, consensus agreement on these subjects. The subject of this book illustrates both the continuing relevance of the concerns of the founders, and how IUPAC contributes to promoting consensus. Equally competent and conscientious labs can arrive at different test results because of their different use of environmental reference matrix reference materials. This can arise from the use of different standards and different procedures. These differences can often be masked by the use of the same or similar terms for different procedures and materials. In the present case the meaning of the different terms Reference Material, Certified Reference Material and Standard Reference Material and the proper use of these materials must be agreed on by the practitioners of the relevant scientific disciplines.

IUPAC's role in achieving consensus is to bring together technical experts so that they can share their experiences and expertise. This often involves other organizations interested in the subject. In this case the co-sponsors of the Workshop, ISO/REMCO, BAM and EUROLAB-D have a vital interest in ensuring that analytical methods using environmental matrix reference materials are consistent among laboratories both within one country, one region and around the world. In our global economy, it is important that test methods used by a laboratory be comparable to those used by laboratories anywhere else in the world.

I congratulate the organizers and contributors to this Workshop on having helped to promote the goal of achieving voluntary consensus through the exchange of technical information. This volume will be an invaluable reference for all those using environmental matrix reference materials in their daily work.

John W. Jost

Foreword

Prof. Dr.-Ing. Dr. h.c. Horst Czichos
President of the Federal Institute for
Materials Research and Testing (BAM)
Chairman of EUROLAB-Deutschland

In broad sectors of analytical chemistry and materials testing, reference materials (RM) are needed to enable laboratories to perform comparable tests and analyses. It is a major task of the Federal Institute for Materials Research and Testing (BAM) - as recently confirmed by the revised pertinent German Federal Act - to develop and to provide reference materials and reference procedures. In this capacity we collaborate in the CCQM (Comité Consultatif pour la Quantité de Matière) of CIPM (Comité Internationale des Poids et Mesures).

In many cases a prerequisite for the development of RM is the co-operation with competent partners. Since many years we are in the fortunate position to collaborate with various organizations and laboratories in the field of metallic RM (e.g. EURO-CRM). Developments in recent years have enabled us to extend our activities to environmental matrix RM. First results of this work are available now. Beyond chemical analysis we provide RM for testing of e.g. porous materials, polymers and surfaces.

To serve trade and industry it is essential to provide information on the specification, availability, applicability and usage of RM. The COMAR database contains information on approximately 10.000 RM from 20 countries. From the very beginning BAM has been one of the COMAR Coding Centres as a founding member and has recently taken over the Central Secretariat from LNE, France.

A keyword for the globalization of trade is "confidence". In the case of testing and analysis this means confidence in the reliability of the results and consequently confidence, too, in the competence of RM producers. As a means to establish confidence accreditation systems have been set-up world-wide. But to make accreditation effective and efficient it was also important that the laboratories have organized in laboratory associations as EURACHEM and EUROLAB. Together with the European Co-operation for Accreditation (EA) these two organizations have set-up jointly the so-called EEE-Group on Reference Materials. In this group not only accreditation procedures and the related documents but also requirements on the quality of RM are discussed.

The development of RM is often time-consuming and complex, in particular in the case of environmental matrix RM. Therefore we need close co-operation between the RM producers to share resources and avoid duplication, and we need feed-back from the RM users to prioritize RM projects. As President of BAM and as Chairman of EUROLAB-D I expect that this international workshop will contribute to this goal.

Horst Czichos

Introduction

This book contains lectures presented at the workshop on "Proper Use of Environmental Matrix Reference Materials", held on April 22 and 23, 1999 in Berlin, Germany. The IUPAC Interdivisional Working Party on Harmonisation of Quality Assurance Schemes for Analytical Laboratories, ISO/REMCO (ISO Committee on Reference Materials), the German Federal Institute of Materials Research and Testing (BAM) and EUROLAB - Germany have co-operated in organisation of this event.

The idea for this workshop originated from an informal discussion at the BERM-7 conference, held in Antwerp, Belgium in April 1997, where many of the problems related to environmental and biological reference materials were highlighted. The characterisation and certification of environmental matrix reference materials is fraught with many complex problems related to traceability and uncertainty of the assigned property values. These are the same materials which are, due to variety of reasons, often misused in the analytical process. Although the analysts are normally instructed on the intended use of the material in the accompanying certificate, the role of matrix reference materials in the analytical process is not always clear. One of the reasons for misuse is the confusion in the nomenclature where similar or the same terms are often used for different types of calibration standards, RM or CRM and internal control samples.

The contributions in this book represent the reference materials producers' point of view of how their reference materials should be properly utilised in an analytical process. A possible application of reference materials is often compared with the guidelines on proper use of reference materials and the requirements of international guides related to reference materials. Worked examples provide additional practical information to the users of reference materials and give guidance for decisions that can be taken on the basis of analytical results obtained on reference materials. Information on future plans and strategies, included in some of the contributions, is also useful for the reader in the sense of having realistic expectations related to availability of SI-traceable natural matrix reference materials in near future, their role in analytical process and as a possible tool for more practical establishment of traceability chains and uncertainty budgets. There are also contributions related to the ISO/REMCO, a central international body dealing with harmonisation and standardisation of topics related to reference materials and the ISO Guide 33, entitled "Proper use of reference materials"

We hope that the practical examples and discussions presented in this book will help the users of reference materials to better understand the quality requirements that reference materials have to fulfil before being selected and applied for a specific purpose in an analytical laboratory. The book, written by reference materials producers, is oriented toward the users - scientists, researchers, technicians and students who use natural matrix reference materials in their daily laboratory work.

Aleš Fajgelj
Michael Parkany

Contents

China GBW Reference Materials

Pan Xiu Rong, Zhao Min

NATIONAL RESEARCH CENTER FOR CRM, BEIJING 100013, CHINA

1 INTRODUCTION

The GBW RMs are the reference materials approved and published by China State Bureau of Technical Supervision (CSBTS) according to China Law on metrology. The mark which stands for China Metrological Certification should be on the certificate and the label of GBW RMs. In order to develop GBW RMs China government made a decision of founding the National Research Center for CRM (NRCCRM) in 1980. The basic mission of the NRCCRM is to establish and realize traceability of measurement in analytical chemistry, physico-chemistry and chemical engineering. Its main tasks are as follows:

- Study primary methods and primary reference materials.
- Research reference methods and certified reference materials.
- Participate in international or bilateral comparison of primary methods and primary reference materials.
- Draft technical documents on RMs and calibration of measuring instruments.
- Perform a certification and accreditation program of GBW RMs and their producers under the authorization of the CSBTS.

Now more than 1800 kinds of GBW RMs are available. They are classified according to metrological level into the primary reference material (PRM), Certified reference material (CRM) and working reference material (WRM), and by application field into 13 categories (Table 1). Most of them have been widely applied to analytical measurements of the below fields:

- Environment monitoring
- Clinical chemistry
- Industrial hygiene
- Food and medicine inspection
- Agriculture chemistry
- Development of materials and energy resources
- Quality management and quality assurance of products
- International and domestic trade.

They provide a foundation of realizing traceability for analytical results and play an important role in the below aspects:

- The development of trace analytical techniques
- The analytical quality assurance
- The accreditation of analytical laboratories

- The arbitrary analysis etc.

In recent years the users of GBW RMs have extended all over the world, such as USA, Canada, UK, France, Germany, Italy, Spain, Sweden, Australia, New Zealand, South Africa, Japan, etc. - about 30 countries.

2 TRACEABILITY OF GBW RMS CERTIFIED FOR CHEMICAL COMPOSITION

Traceability is defined as the property of the result of a measurement or the value of a standard thereby it can be related to a stated reference, usually a national or international standard, through an unbroken chain of comparisons having stated uncertainties[1]. This definition implies that traceability is an essential property of a measurement result and measurement standard; the traceable result of analytical measurement must be an

Table 1 *Summary of Available GBW RMs*

Category	Grade and Amount			Total
	PRM	CRM	WRM	
Environment Chemistry	58	63	343	464
Geology and Ores	-	213	63	276
Physico-Chemistry	35	35	156	226
Ferrous Metals	-	211	30	241
Non-Ferrous Metals	3	141	10	154
Nuclear and Radioactivity	1	112	11	124
Chemical Industry	10	18	153	181
Clinical Chemistry	4	31	17	52
Food Chemistry	-	5	8	13
Building Materials	-	35	2	37
Energy Resources	-	25	18	43
Technology and Engineering	3	1	12	16
Polymer	-	2	3	5
Total	114	892	826	1832

interval covering true value of the measurand and is expressed by $\overline{X} \pm u$. If an analytical measurement is performed against a suitable and traceable RM, and analytical process is in a state of statistical control, it is possible to trace the analytical result back to the SI units. However, it is very difficult to establish the traceability for RMs certified chemical composition. In order to promote the establishment of traceability for GBW RMs the CSTBS published the following documents from 1982 to 1987:
- Means of examining and authorizing to produce GBW RMs.
- Method of Numbering GBW RMs.
- Regulation of compiling certificate of GBW RMs.
- General Terms of GBW RMs.
- Technical Norm of the preparation and certification for GBW RMs.
- Rule of Administration for GBW RMs.

The certification and accreditation program for GBW RMs grades and their producers was started by NRCCRM in 1983. The national measurement system of analytical measurement is being formed. In China more and more chemical metrologists and analytical chemists pay a attention to the traceability of GBW RMs and make efforts for it.

2.1 Traceability mode and chain of analytical measurement

The analytical measurements have become an important part of human activities. The results of chemical measurements are often used to make decisions relating to economy, technology and legislation. Therefore the traceability of analytical results are required. It is clear that the realizing traceability for analytical measurement can not follow the traceability mode and chain of physical measurement due to the particular problem and matter of analytical measurement. Many international guidelines and documents [1-12] regarding analytical quality assurance, analytical laboratory accreditation, methodology, reference material, calibration and uncertainty , etc. have been issued. These should facilitate the establishment of international traceability in analytical measurement. The interrelationships and roles of them in establishing traceability of analytical measurement are described in figure 1. The four possible modes of realizing traceability are shown by Fig 1. It must be pointed out that uncertainty and inter-comparison are the key factors in all traceable approaches, but Fig 1 does not indicate them.

In fact, these possible traceability modes have different practicability and suitability for analytical measurement. Mode D has been successfully realized for physical measurement, but is not suitable for chemical measurement. Mode C is suitable, but is limited by space, time and fields. Mode B has more difficulties than mode A in practice because there are a few primary methods and reference methods in analytical measurement, and it is more difficult for field laboratories to use them. A variety of RMs are available and are widely used in chemical measurement, and it is easier for field laboratories to get and use RMs, so that mode A has better practicability and suitability for establishing traceability in analytical measurement. The main problem to be solved for mode A is how to ensure traceability of certified values of RMs . The author thinks that the best way to solve this problem is to establish a traceability chain as in Figure 2 for RMs or analytical measurement.

A traceability chain as in Fig.2 has three features which are suitable for all measurements using RMs as measuring standards, and are flexible, enabling field laboratories to select different grades of RMs to meet the requirements of field analysis. There are two paths, the right-side path is suitable for measurable quantities relating to the chemical composition of the natural matrix substance, and the left-side path is for quantities relating to the chemical composition of a solution or gas mixture. Figure 2 indicates that the traceability chain is formed of eight links from field analysis to SI units, and both RMs and methods should be classified into three grades. These grades of RMs and methods have been defined by the international organizations concerned.

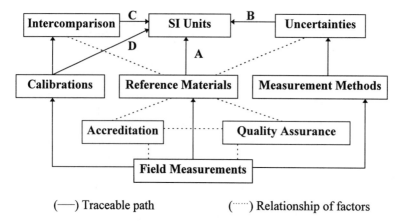

(——) Traceable path   (······) Relationship of factors

Figure 1 *Block diagram of traceability mode*

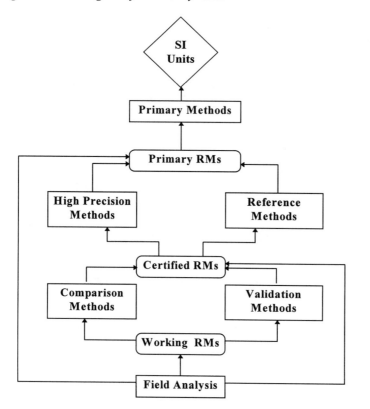

Figure 2 *Traceability chain for analytical measurements*

Primary method

A primary method of measurement is a method having the highest metrological qualities, whose operation can be completely described and understood, for which a complete uncertainty statement can be written down in terms of SI units, and whose results are, therefore, accepted without reference to a standard of the quantity being measured [2].

Primary RM

A primary RM is one having the highest metrological qualities and whose value is determined by means of a primary method [2].

Reference method

This is a thoroughly investigated method in which exact and clear descriptions of the necessary conditions and procedures are given for the accurate determination of one or more property values; the documented accuracy and precision of the method are commensurate with the method's use for assessing the accuracy of other methods, for measuring the same property values, or for assigning reference values to reference materials [3].

Certified RM

RM, accompanied by a certificate, one or more of whose property values are certified by a procedure which establishes traceability to an accurate realization of the unit in which the property values are expressed, and for which each certified value is accompanied by an uncertainty at a stated level of confidence [3].

Validation method

An analytical method which has been validated by systematic laboratory studies, and whose performance characteristics meet the specifications relating to the intended use of the analytical results. The performance characteristics determined include selectivity and specificity, range and linearity, limit of detection and quantification, accuracy and precision, ruggedness etc. These parameters should be clearly stated in the documented method [4].

Working RM

RM which is used routinely in field analysis for calibrating an analytical process or instrument, for assessing an analytical method, or for assigning values to a substance analyzed [1].

The SI units in Fig 3 for the quantities relating to the chemical composition mainly comprise: mole, kilogram and sub-multiples of mole and kilogram; mole fraction, mass fraction; derived units of concentration mol/m^3, kg/m^3, mol/kg; Faraday constant, gas constant, molar mass etc. In order to establish international traceability for quantity values of chemical compositions the CIPM/CCQM has initiated international comparison studies on the primary methods and primary RMs between national metrology laboratories.

2.2 The certification and accreditation of GBW RMs grade and their producer

The certification and accreditation of grades and producers of more than 1800 GBW RMs have been completed in accordance with criteria of the table 2 and program shown in figure 3 since 1983. The facts demonstrate that the certification and accreditation program of GBW RMs plays an important role in facilitating establishment traceability for GBW RMs.

2.3 Chinese national measurement system for analytical measurement

The national measurement system for analytical measurement is being formed in China. Its situation is illustrated in the figure 4.The CSBTS in Fig 4 is a chinese government agency for metrology, standardization and quality. The NRCCRM belongs to it and bears the responsibility for the secretariat and chairman of three committees, i.e. Reference Materials Committee(REMCO), Chemical Metrology Committee (CMCO), and Analytical Quality Assurance Committee (AQACO). The members of each committee are the representatives of institutes, academies and producers concerned, and are the experts on analytical chemistry ,chemical metrology , and quality assurance from the whole country . They often hold symposia, workshops, and seminars to promote the establishment of the traceability, to provide wide contact with the laboratories of field analysis, and to give advice on chemical measurements. The NRCCRM often partake in international and bilateral comparison to ensure the international equivalency of GBW RMs. ,A number of analytical laboratories of institute and academies, as reference laboratories, are engaged in research of analytical methods and RMs or in calibration service for analytical instruments. The laboratories for routine analysis select and use suitable RMs and do analytical measurement according to their quality assurance program.

Table 2. *The Grade Criteria of GBW RMs*

Grade	Name	Criteria
I	Primary reference material (PRM)	• Satisfy the definition of a PRM • be developed by a national metrology laboratory • be certified by a primary methods • be recognized by a national decision • can be traced back to SI units and verified by international comparisons.
II	Certified reference material (CRM)	• satisfy the definitions of a CRM, of a reference standard or of a secondary standard [1] • be usually developed by a national (or specialized) reference laboratory • be certified by reference methods or high precision methods, or by a combination of different methods • be recognized by national (or specialized) authoritative organizations • be accompanied by a complete statement of demonstrated uncertainty and traceability.

Grade	Name	Criteria
III	Working reference material (WRM)	• satisfy the definitions of an RM and of a working standard [1] • be produced by an accredited body • be certified by the specified validation methods • be accompanied by a clear statement of uncertainty and evidence of traceability.

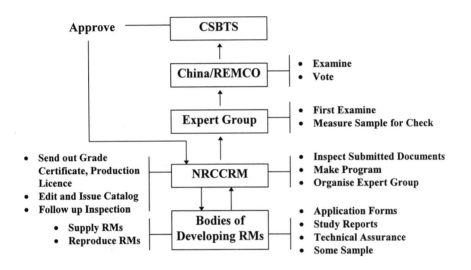

Figure 3 *Certification and accreditation scheme for GBW RMs grade and their producer*

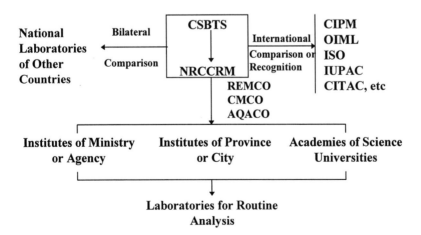

Figure 4. *Chinese national measurement system for analytical measurement*

3 PROCEDURE AND KEY POINTS OF RESEARCH GBW RMS

It is very necessary to focus all research work on the traceability of certified values of RM, i.e. the purpose of all work is to reasonably assign the property values and their uncertainties of RM. The general procedure of research GBW RMs is summarized in the table 3. The key points are described as follows.

Table 3 *General Procedure of Research of GBW RMs*

Procedure	Content	Principles and criteria
Determine technical target	1.investigate where the RM will be used, what is the practical requests. 2.determine the type of material and technical target. 3.determine the range of property value and uncertainty to be certified	advanced techniques, reasonable economy, good applicability and suitability
Tentative study	1.study the homogeneity and stability of candidates. 2.determine the range of property value to be certified. 3.choose suitable material from candidates.	homogeneity stability can meet practical request. The range of property value is expectant.
Select certification methods	1.primary methods, performed by a national metrological laboratory. 2.two or more accurate and reliable methods, carried out by authorized laboratories. 3.interlaboratory testing. 4.combined methods 2. and 3.	method 1. is the best one, it should be used as much as possible ; method 2. a few laboratories have ability and condition to use; method 3. is common technique; method 4. is more economic and reasonable one than method3.
Study analytical methods	1.study the accuracy of measurement methods including recovery, linearity analysis blank , precision, bias (between methods or laboratories) 2.study precision of method used in homogeneity and stability tests.	1.the uncertainty of measurement method should be better than uncertainty of certified value of RM. 2.the more precise the more good, at least it can test the un-negligible variation of material.

Procedure	Content	Principles and criteria
Collect and prepare sample	1. collect enough material, pay attention preventing it from contamination and deterioration. 2.select scientific and reasonable procedure to prepare homogeneous and stable sample in batches. 3.pack the sample in units which are suitable for use, storage and transportation, 4.test homogeneity and stability, provide minimum sample size and expiry date	1. the amount should be enough for use in valid period 2. the error of heterogeneity can be neglected compared with random error of measuring method, or give out the estimation of heterogeneity. 3.the stability of material is not aware change within stipulated period. 4.package must ensure that no changes of property value occur during transportation and storage.
Certify	1.work out a certification plan based on homogeneity testing and selected certifying approach which include number of laboratories or methods, number of sample to be measured and replicate times. 2.determine the property value using the methods studied. 3.check if there are outliers for technical reasons, dispose the outliers.	1. measure more samples of material which heterogeneity is not negligible, or measure them individually; 2. measure 2-3 samples by one laboratory or one method, each sample must be measured twice when heterogeneity is negligible. 3. for inter-laboratory testing the number of laboratory should be not less than 8, the number of laboratory can be reduced when more methods are used. 4. it is better to use known standard similar to sample to be measured and to measure blank for quality control.
Estimate certified value and uncertainty	1. collect all data, make statistical analysis, 2. synthetically analyze homogeneity and stability, estimate the variation of material, 3. give the optimum evaluated value of property certified, namely certified value and uncertainty.	1. determine the certified value and its uncertainty according to the data analysis, the reliability of method, variation of material and possible bias between methods or laboratories.
Comparison test and probation	1.compare this material with a similar RM produced by other producer. 2.ask some users to use it.	good agreement, no problems in use

Procedure	*Content*	*Principles and criteria*
Draw up certificate	write down all information of the RM briefly and clearly, introduce the property values and application of RM, give a guarantee on reliability of certified values	1.users can get ideas about the reliability of certified value 2.give a guidance of use and store of RM.
Issue to use	1. apply for a "license of manufacture" and "grade certificate" 2.officially issue to the public. 3.test stability continuously 4.reproduce RMs.	guarantee the reliability and validity of certified value

3.1 Homogeneity assurance of RM

The materials such as mixed powders, ores, alloys are naturally heterogeneous in composition. The so called homogeneity of materials is the description of identity of one or more specified properties. Theoretically material is homogeneous with respect to a given property if there is no difference between the values of this property from one part to another. In practice, how to know there is any difference? The common way is experimental examination. If a difference between the value of a specified property from one part to another can not be detected, then material is accepted to be homogeneous with respect to the specified property. The practical concept of homogeneity therefore embodies both a specificity to the characteristic and a parameter of measurement (usually the standard deviation) of the measurement method used, including the defined sample size of test portion. For same properties of samples, a different conclusion of homogeneity may be obtained, because different sample size was taken or different methods were used in the examination. In a word, homogeneity of material is the description state of specified property of material in different portion, it may distribute random or have certain tendency.

3.1.1 Test of homogeneity. The RMs in solid and powder should subject preliminary homogeneity test during preparation and after subdividing into units. Even for liquid and gas, it is also necessary. The main purpose of this test is to find out unexpected problems happened in preparation and packing, such as contamination, un-complete dissolution and equilibrium (which could lead to change of concentration from the fist vial filled to the last).

Homogeneity test must be performed for the candidate of RM after it was packed into final form to confirm whether the between-units variation is statistical or practically significant. For homogeneity test, a well-conceived plan should be devised; it may consist of :

1) Decide sampling mode, systematic or random sampling based on the preliminary understand of the material;

2) Decide the number of samples, which should be enough to represent the degree of homogeneity of the whole batch material;

3) Decide sampling times of within-unit to the number of replicate measurement;

4) Decide sample size, which should be commensurate with the requirement of practical measurement;

5) Choose representative property for test when material has more certified properties;

6) Select measuring method of high precision, repeatability and reproducibility. The higher the degree of homogeneity of material is ,the more precise a measuring method must be employed;

7) Decide laboratories in perfect working conditions, choose equipment with good metrological performance and assign professional persons for homogeneity testing.

3.1.2 Evaluation of homogeneity. The homogeneity evaluation is based on great amount of measuring data, through comparison of variance to determine if significant difference of property values exist.

1) Test within-unit sub-samples and evaluate if there is significant difference between sub-samples. First, to determine the variance of measuring method, S_a^2; second, to measure several sub-samples taken from one unit, calculate the variance of the data, S_0^2. In fact S_0^2 is the sum of S_a^2 (measuring method) and S_s^2 (sub-samples) ,i.e. $S_0^2 = S_a^2 + S_s^2$. If no significant difference is found between sub-samples, the S_s^2 could be neglected. And S_0^2 approximately is equal to S_a^2. This means that material is homogeneous within-unit.

2) Test between-units samples and evaluate if there is significant difference between samples taken from different units.

Suppose m units are taken out of a lot of RM , n sub-samples are taken from each unit, measure these samples and get the following data:

$$
\begin{array}{llll}
X_{11}, X_{12}, X_{13}\ldots\ldots\ldots X_{1n}, & \overline{X}_1 \\
X_{21}, X_{22}, X_{23}\ldots\ldots\ldots X_{2n}, & \overline{X}_2 \\
\ldots, \ldots, \ldots\ldots\ldots\ldots\ldots, & \ldots \\
X_{m1}, X_{m2}, X_{m3}\ldots\ldots X_{mn}, & \overline{X}_m
\end{array}
$$

Calculate overall mean $\overline{\overline{X}}$ and sum of variances of Q_1 between units, and Q_2 within-unit, the degree of freedom f_1 and f_2 with the following equations:

$$\overline{\overline{X}} = \frac{1}{m}\sum_{i=1}^{m}\overline{X}_i$$

$$Q_1 = n\sum_{i=1}^{m}\left(\overline{X}_i - \overline{\overline{X}}\right)^2$$

$$Q_2 = \sum_{i=1}^{m}\sum_{j=1}^{n}\left(X_{ij} - \overline{X}_i\right)^2$$

$$f_1 = m - 1$$

$$f_2 = m(n-1)$$

compare $\dfrac{Q_1}{f_1} / \dfrac{Q_2}{f_2}$ with table value of F distribution at a given significant level. If the former is less than the later. It can be recognized that there is no significant difference between within-unit and between-units at a given significant level.

3) Test and evaluate the difference between-units

When it is impossible to withdraw many sub-samples from one unit, then take out m units from a lot of RM, measure them and get m data, calculate variance S_o^2, compare S_o^2 with the variance of measuring method S_a^2, if S_o^2 / S_a^2 is less than the value of $F_{0.05}$ table, then it is recognized that there is no significant difference between-units of material, i.e. the material is homogeneous.

The property value of materials in rod, wire or piece form is usually related to their geometrical positions. When evaluating the homogeneity of these kind of materials, an appropriate mathematical expression mode should be established first.

3.2 Stability assurance of RMs

The chemical composition and physico-chemistry properties of materials may change with time, because of the effect of physico-chemical and biological factor. The stability of RM is an ability, by which the variability of stated property value can be kept in a given uncertainty under condition of storage and within a certain time interval. The following works should be done when study stability of RM:

1) According to the property of material to be studied, an appropriate preparing procedure and necessary technical conditions should be taken to pretest the stability. For example, human serum and urine RMs are usually prepared with vacuum freezing-dry method under germ-free condition, UV light or ^{60}Co -ray are used to irradiate RMs after subdivided into final units to kill germ. Acids are usually used as stabilizer for metal elements in water, etc.

2) Select practicable condition for storage of RMs. As we know low temperature and humidity are suitable for biological RMs. Containers filled with inert gases are proper for RMs which are easily oxide. Since some conditions cost a lot and are not convenience for transportation and use of RMs, this is only one of the choice when it has to do. Generally a feasible condition should be selected through experiments. Container affects the stability of RMs strongly, sometimes it may become a decisive factor. Therefore it's better to use containers which made of suitable materials and confirm whether their absorptivity, permeability, solubility and sealability etc. could meet the requests of RMs storage.

3) Examine the change of the property value with time. The sampling method, the number of unit and the representative property to be measured are determined based on the factors which influence the stability, the differences of stability of the properties and the amount of RM prepared. The standard deviation of the method used for property value examining should be less than 1/3 the expectant value of uncertainty for property value, or at least as same as the precision of the method which is one of the most precision methods used for certification. In order to get the pattern of variation of property value, a suitable time interval of testing must be set. The sample for examination should be not less than 7 and each sample should be measured one time at least. A check sample for quality control may be used during the measurements . The mean value $\overline{X}$ and standard deviation S_i then can be calculated from these data. Take $\overline{X}$ as ordinate, the related time as abscissa, a variation graph of property value/time then is plotted.

4) Evaluate stability of RM. There are 5 possibilities of variation graph of property value of RM *vs*. time: the property value increases or decreases continuously along with time; the property value increases or decreases continuously just after the RM was prepared, a period of time later, it tends to smooth and last for sometime; the property value is stable after RM was prepared, but it increases or decreases later; the property value increases or

decreases slowly along with time; the property value vary randomly. For the first possibility the stability of the property value is very bad. In this case, it is necessary to improve the stability or select other material to instead. For the 5th possibility ,it requires to analyze and adjudge using Cochran test if all measurements are equal precision, calculate the pool standard deviation $S_p = \left(\sum S_i^2 / m\right)^{1/2}$. Si is experimental standard deviation of n sample, m is stability testing times. If 95% values obtained fall randomly into the range of $\overline{\overline{X}} \pm 2S_p$, the variance of property value is due to the random error of measurement. The variation of the property value with time was not detected. For the 2nd, 3rd and 4th possibilities the feature that the property value changed in certain pattern and at slow speed may help us to select strict conditions and determine the valid time of the RM through detecting the variation of the property value with time.

5) Determine valid period. The valid period of RM is often determined by its producer through experiment. The valid period means the period from the day on which RM was certified to the final day after which RM can't be used anymore, sometime the valid period only means the storage time. Some RMs can be only used one time. When the package is opened it must be used at once. Mostly the valid period means the use period under certain conditions. Since the valid period is related to preparing techniques, storing conditions and utilization, it should be determined by experiment for each preparation. For RMs having much longer valid period, a tentative use period could be given to them based on the data obtained in the past. As testing it continuously the use period may be extended. When the change of the property value is out of the range of uncertainty, then the user must be notified to stop use.

3.3 Approaches to certifying property values of RMs

The certified value of RM must be the best estimate value of "true value". It is the most important to study and use measurement methods having high metrological qualities as possible. During the certification of RMs the quality assurance program should be performed, such as the metrological calibration of measuring instruments, process control of measurement, assessing accuracy of measurement results etc. The main approaches to certifying property values of GBW RMs are as follows:

1) A single primary method of measurement by a single national metrological laboratory. For instance, NRCCRM have used the constant-current coulometry, controlled-potential coulometry, gravimetry, isotope dilution with TIMS, isotope dilution with GC/MS, determination of freezing-point depression etc. primary methods to certify property values of primary RMs.

Table 4 *Some Natural Matrix GBW RMs*

Code number	Matrix	Code number	Matrix
GBW08501	Peach leaves	GBW08571	Mussel
GBW08504	Cabbage	GBW08572	Prawn
GBW07602	Bush branches and leaves	GBW08551	Swine liver
GBW07604	Poplar	GBW09101	Human hair power

Code number	Matrix	Code number	Matrix
GBW07605	Tea	GBW09102	Lyophilized human urine
GBW08502	Rice flour	GBW09104	Lead in human urine
GBW08503	wheat flour	GBW09106	Fluoride in human urine
GBW08507	Corn flour	GBW09131	Bovine serum
GBW08508	Rice flour	GBW09132	Bovine blood
		GBW09135	Human serum

Table 5 *Summary of Some Natural Matrix GBW RMs*

Element	Content Range (μg/g)	Relative Uncertainty (%)	Analytical method
Ag	0.03-0.05	20-10	AAS,ICP/MS,NAA.
Al	13-2*	18-9	AAS,ICP,NAA,SP, XRF,ISE.
As	0.05-6.0	20-18	AAS,AFS,ICP/MS, NAA,POL,SP,SF,ASV.
B	15-50	17-8	ICP,ICP/MS,ISE.
Ba	4.3-60	17-5	ASS,ICP,ICP/MS, IDMS,NAA,XRF.
Be	0.02-0.06	20	AAS,ICP,ICP/MS.
Bi	0.02-0.06	18-11	AFS,ICP/MS.
Br	2-7	12	IC,NAA,SP,XRF.
Ca	85-2*	7-3	AAS,AFS,ICP,NAA, PIXE,XRF.
Cd	0.03-4.5	13-10	IDMS,NAA,POL,IC.
Ce	0.1-2.5	20-8	AAS,ICP/MS,IDMS, POL.
Cl	150-1.9*	6-4	IC,IDMS,NAA,SP,VOL.
Co	0.07-0.95	12-7	AAS,ICP,ICP/MS,NAA.
Cr	0.37-4.8	14-8	AAS,ICP,ICP/MS,NAA,PIXE.
Cs	0.05-0.29	7	ICP/MS,NAA
Cu	0.05-23	20-6	AAS,ICP,ICP/MS,IDMS,PIXE,POL,XRF,ASV.
Eu	0.009-0.037	22-5	ICP,ICP/MS,NAA.
F	0.62-320	7-9	IC,ISE,SP.
Fe	5.2-0.105*	13-8	AAS,ICP,NAA,XRF.
Hg	0.026-2.2	12-10	AAS,AFS,MIP,NAA,SF.
K	219-1.97*	10-7	AAS,IC,ICP,ISE,NAA, XRF,IC.
La	0.049-1.25	17-3	ICP,ICP/MS,IDMS, NAA.
Li	0.84-2.6	13-11	AAS,ICP,ICP/MS,FP.
Mg	20-0.65*	6-5	AAS,ICP,ICP/AFS, NAA,XRF.

Element	Content Range (µg/g)	Relative Uncertainty (%)	Analytical method
Mn	0.28-1240	18-3	AAS,ICP,ICP/MS,ICP/AFS,NAA,PIXE,XRF.
Mo	0.038-3.8	16-20	AAS,ICP/MS,IDMS,NAA,POL.
N	1.20*-14.9*	2-0.7	kje,SP,VOL,IC.
Na	44-1.96*	9-5	AAS,AES,ICP,NAA,XRF,IC.
Nd	1.0	10	ICP,ICP/MS,NAA.
Ni	0.83-7.6	18-7	AAS,ICP,ICP/MS,ICP/AFS,IDMS,NAA,PIXE.
P	17.0-4.88*	4-3	ICP,POL,SP,XRF,IC.
Pb	0.54-8.8	8-10	AAS,ICP,ICP/MS,IDMS,POL,PIXE,XRF,ASV.
Rb	4.2-74	5-6	AAS,ICP/MS,IDMS,NAA,XRF.
S	0.25-4.3*	6-5	IC,ICP,SP,VOL,XRF.
Sb	0.037-0.095	8-13	AFS,NAA,POL.
Sc	0.008-0.32	13-9	ICP,ICP/MS,NAA.
Se	0.04-0.94	25-6	AAS,AFS,NAA,MFS,POL,SP,IGC,SF.
Si	87-0.60*	8-6	GRA,SP,XRF.
Sm	0.038-0.19	11-6	ICP,ICP/MS,NAA.
Sn	0.12-0.23	-	ICP/MS,AES,POL.
Sr	4.2-345	4-2	AAS,ICP,ICP/MS,IDMS,NAA,PIXE,XRF.
Tb	0.025	8	ICP,ICP/MS,NAA.
Th	0.061-0.37	13-6	ICP/MS,IDMS,NAA.
Ti	2.7-95	15-14	AAS,ICP,SP,XRF.
V	2.4	9	ICP,ICP/MS,NAA,POL,XRF.
Y	0.084-0.68	19-3	ICP,ICP/MS.
Yb	0.018-0.063	17	ICP,ICP/MS,NAA.
Zn	0.60-198	15-5	AAS,ICP/MS,ICP,ICP/AFS,IDMS,POL,NAA,PIXE,XRF.

* content in wt%.

Note for analytical methods:

AAS:	Atomic Absorption Spectrometry
AES:	Atomic Emission Spectrometry
AFS:	Atomic Fluorescence Spectrometry
ASV:	Anodic Stripping Volumetry
DNA:	Delay Neutron Assay
FP:	Flame Photometry
GRA:	Gravimetry

GRS: Gamma-Ray Spectrometry
IC: Ion Chromatography
ICP: Inductively Coupled Plasma Spectrometry
ICP/MS: ICP Mass Spectrometry
ICP/AFS: ICP Atomic fluorescence Spectrometry
IDMS: Isotope Dilution Mass Spectrometry
IGC: Inorganic Gas Chromatography
ISE: Ion Selective Electrode
kje: Kjeldahl Method
LFS: Laser Fluorescence Spectrometry
MFS: Micro Fluorescence Spectrometry
MS: Internal Standard Spark Source Mass Spectrometry
MIP: Microwave Induced Plasma Spectrometry
NAA: Neutron Activation Analysis
POL: Polarography
PIXE: Proton Induced X-ray Emission Analysis
PK: Photochemical Kinetics
SF: Spectrofluorimetry
SP: Spectrophotometry
VOL: Volumetry
XRF: X-ray Fluorescence Spectrometry

2) Two or more independent reference methods by a single authorized laboratory such as NRCCRM.

3) Two or more independent validation methods by several laboratories which have equal competence for certifying stated RM. This approach is often applied to certifying content of trace composition in natural matrix RMs. some natural matrix GBW RMs and their outline are gave in the table 4 and 5.

3.4 Evaluation of uncertainty for certified value of RMs

The uncertainty of certified value of RM is very important for establishing traceability of RM and to realize comparability of analytical results. Evaluating uncertainty of certified value of GBW RM is not only basing on the measured data and the assessment of homogeneity and stability of RM, but also depending on the experience justification of the expert. In practice the differences between measuring methods or laboratories are not avoided. Thus, the uncertainties arising from them should be counted. Besides, analytical measurement is complex procedure which includes eight main sources of uncertainty from sampling to reporting the result shown in Fig 5. They are the uncertainties concerned with homogeneity (u_1), recovery(u_2), analysis blank (u_3),measurement standard (u_4), calibration (u_5), matrix effect and interference (u_6), measuring instrument (u_7) and data processing (u_8) respectively. In addition, bias effects may occur in the sampling step, treating the sample, and carrying out the measurement. They are expressed by T_1-T_0, T_2-T_1, and $\overline{X} - T_2$ and are caused by the difference between the sample and population, the loss and contamination of analyte, and the difference between the sample matrix and that of the measurement standard (i.e. the RM) respectively. Before evaluating uncertainty, the work of finding and correcting bias has to be done. The uncertainties arising from these corrections should also be counted.

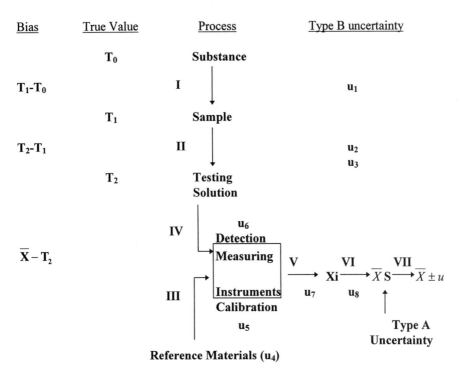

Figure 5 *Typical analytical measurement and its uncertainty*

The combined standard uncertainty (u_c) of certified values of GBW RMs is expressed by the below equation according to international guide of uncertainty in measurement [5].

$$u_c = \left(\sum_i S_i^2 + \sum_j u_j^2 + \sum_k u_k^2 \right)^{1/2}$$

where

S_i^2 Variances arising from repeated measurements and assessment of homogeneity and stability.

u_j Type B standard uncertainties arising from the measurement standards and constants cited, and the corrections of possible systematic errors and others in the analytical procedure.

u_k Type B standard uncertainties based on experiences concerned with different methods and laboratories.

Sometimes some components of uncertainty can be neglected. So that one component of them can approximately express u_c. In this case the repeatability or reproducibility of results of measurements are often a representative of uncertainty of certified value of RM.

4 EXAMPLES OF GBW RMS CERTIFIED FOR CHEMICAL COMPOSITION

Looking over the GBW RMs catalog of China, it can be found that 10 of 13 categories, about 1300 kind of PRMs, CRMs and WRMs, are relating to the analysis of chemical compositions. For convenience of illustration, they may be divided into 4 types: high-pure (inorganic and organic) substance; (Single and multi-components) solution; matrix CRMs and gas mixture.

4.1 GBW RMs certified for purity of substance

There are 3 ways of certifying purity of a substance, one is measuring the main composition, the other is measuring the impurities, and the third is measuring both. Table 6 shows some examples of GBW RMs whose purities were certified by measuring the main composition.

The RMs were prepared specially ,the purities were certified with high-precision coulometry which is based on Faraday's law of electrolysis. This is a primary method for stoichiometric purity determination. The principal equation is $m = \dfrac{M}{ZF} \cdot It$, (m: mass of reaction substance; M: molar mass of reaction substance; F: Faraday constant; ; Z: number of reaction electron; I: electric current; t: reaction time). For a given substance M/ZF is constant, I and t can be measured accurately . If know the mass (m') of sample introduced into the electrolytic cell, the purity of substance being measured could be calculated and expressed with mass fraction (m/m'). In order to ensure the accuracy of the certified value, the current efficiency and the error of the end point were studied, the uncertainty of time and current measurement, sample weighing, molar mass and Faraday constant were evaluated.

Table 6 *GBW PRMs of High-pure Substance*

Code number	Name of RM	Certified value and uncertainty (%)	Usage
GBW06101a	Sodium carbonate	99.995 ± 0.008	basimetric value
GBW06102	Na-EDTA·2H$_2$O	99.979 ± 0.005	complex value
GBW06103	Sodium chloride	99.995 ± 0.005	assay standard for Cl$^-$
GBW06105a	Potassium dichromate	99.990 ± 0.003	oxidimetric value
GBW06106	Arsenic trioxide	99.963 ± 0.003	reductometric value
GBW06107	Sodium oxalate	99.960 ± 0.008	reductometric value
GBW06108	Acid potassium phthalate	99.998 ± 0.007	acidimetric value
GBW02751	Gold	99.994 ± 0.037	assay standard for Au
GBW02752	Silver	99.994 ± 0.044	assay standard for Ag

The uncertainty of certified value consists of type A(experimental standard deviation of the mean, $\dfrac{S}{\sqrt{n}}$) and type B (uncertainties of measuring physical quantities and fundamental constants) standard uncertainties. The same method was also used in testing homogeneity and stability.

Another example is chlorine-containing pesticides and their metabolite P.P'-DDT, O.P-DDT, P.P'-DDD and P.P'-DDE. The 4 substances were purified by re-crystal procedure and the purity was measured by two methods: high-performance liquid chromatography (HPLC/UV detection, normalization method, for main composition) and differential scanning calorimetry (DSC, for total impurities). The uncertainty of the certified value is estimated based on the uncertainty of the two methods, the results are shown in table 7.

Table 7 *Certified Value of Chlorine-containing Pesticides*

Code number	Substance	Certified value	uncertainty (%)
GBW(E)06405	P.P-DDT	100.0	0.3
GBW(E)06406	O.P-DDT	99.5	0.3
GBW(E)06407	P.P'-DDD	99.7	0.3
GBW(E)06408	P.P'-DDE	99.9	0.3

4.2 GBW RMs certified for concentration of solution

GBW RMs solution of metal elements were prepared by gravimetric method. Figure 6 shows the procedure. In order to calculate
accurately concentration (c_1) many effecting factors such as matrix, contamination, container influence, stability, homogeneity etc. were studied. The primary method (coulometry) was used to assay the stock solution. Some analytical methods were also used to determine the concentration (c_2) and test homogeneity and stability of the prepared solution. One example is Cd, Cr, Cu, Ni, Pb and Zn in simulation natural water which is used in environmental monitoring. The concentrations of these elements were determined by atomic absorption spectrometry (AAS), inductively coupled plasma emission spectrometry (ICP) and polarography (OL) etc. The analyzed results were identical with that of preparation. The certified values and uncertainties were given based on the measurement and preparation. Table 8 shows some of them.

Another example is metal element in water used as stock solution.

Table 8 *PRMs of Metal Elements in Simulation Water*

Code Number	GBW08607(μg/g)		GBW08608(ng/g)	
	certified value	uncertainty	certified value	uncertainty
Cd	0.102	0.002	10.2	0.4
Cr	0.512	0.010	34	2
Cu	1.02	0.01	51	2
N	0.512	0.005	61	2
Pb	1.02	0.02	51	2
Zn	5.10	0.05	92	4

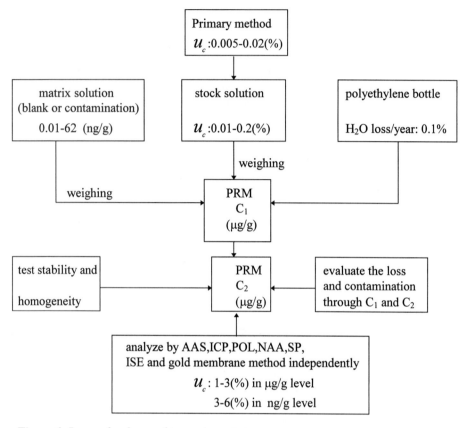

Figure 6 *Research scheme of trace elements in water*

Table 9 *PRMs of Metal in Water*

Code number	Element	Concentration ($\mu g/mL$, 20 $^{\circ}C$)	Matrix
GBW08610	Ag	1000 ± 1	1% HNO_3
GBW08611	As	1000 ± 1	1% HNO_3
GBW08612	Cd	1000 ± 1	1% HNO_3
GBW08613	Co	1000 ± 1	1% HNO_3
GBW08614	Cr	1000 ± 1	1% HNO_3
GBW08615	Cu	1000 ± 1	1% HNO_3
GBW08616	Fe	1000 ± 1	5% HNO_3
GBW08617	Hg	1000 ± 1	3% HNO_3
GBW08618	Ni	1000 ± 1	1% HNO_3
GBW08619	Pb	1000 ± 1	1% HNO_3
GBW08620	Zn	1000 ± 1	1% HNO_3

solution .The concentration of them is 1000g/mL. The matrix is 1%-5% HNO_3 and HCl.
The concentration was certified using a controlled-potential coulometry. Some of them are

tabulated as shown in Table 9.

GBW(E)060133 is a solution of organic composition. The solution contents 8 components i.e. -BHC,-BHC, -BHC, -BHC, P,P'-DDT, O.P'-DDT, P.P'-DDD and P.P'-DDE. The concentration of each component is 0.1mg/mL,the uncertainty is 2.0%. The solution was prepared with CRM (GBW06401-6404) BHC, (GBW06405-06408) DDT and high purity methanol using gravimetric method. The concentration value is taken as certified value, the uncertainties of preparation and GC testing were considered into uncertainty of certified value.

4.3 Matrix RM

Matrix RMs are very important in clinical, environmental, food and geological analysis. There are many GBW matrix RMs. The following 3 examples may present the basic situation of them.

Example 1: GBW08509 elements on non-fat milk powder (Table 10).

The non-fat milk powder was vacuum dried and screened (80-100 mesh), then packed into glass bottles under nitrogen atmosphere. K, Na, Ca, Zn, Cl, Mn and Se were measured for homogeneity testing by X-Ray fluorescence spectrometry (XRF), instrumental neutron activation analysis (INAA), inductively coupled plasma emission spectrometry (ICP) and micro-fluorescence spectrometry (MFS). The stability was evaluated by measuring Fe, Se, Zn, N and Ca using ICP, MFS, POL, Sp and IC methods.

Table 10 *Certified Value of Elements in Non-fat Milk Powder*

Element	Certified value uncertainty	Certifying method	Element	Certified value uncertainty	Certifying method
K	18.6 ± 0.5	a.b.c	Zn	46.8 ± 2.8	a.d.g.i
Na	5.26 ± 0.17	a.b.c	Mn	0.28 ± 0.05	b.d.g.h
Ca	12.2 ± 0.5	b.c.d	Pb	0.034 ± 0.010	a.g.i
Mg	1.30 ± 0.05	a.c.d	Fe	5.18 ± 0.73	a.b.d
N	55.1 ± 2.3	e.e.f	Se	0.22 ± 0.02	j.b.h
P	10.7 ± 0.3	c.e.d	As	0.031 ± 0.011	d.k.i
Cl	11.2 ± 0.4	b.c.e	Rb	(18)	b
S	(3.33)	(c)	Br	(15)	b
Cu	0.26 ± 0.05	c.d.g.i	Hg	(0.0005)	a.m.n

Note: a: AAS   b: INAA   c: IC   d: ICP   e: SP
          f: KMN   g: ICP/MS h: POL   i: DPASR j: MFS
          k: AFS   l: ASV   m: CAS   n: CAF

The content of the elements were certified by several laboratories using more than 2 methods. The average of measured value obtained by each method and laboratory was taken as certified value for each element, two times of standard deviation was taken as uncertainty.

Example 2. GBW08514 and 08515, CRM for composition analysis of tobacco.

There are two tobacco CRMs, GBW08514 is sun-curde tobacco, GBW08515 is flue-cured tobacco, 14 components and elements were certified (by 3 laboratories) using two or more reliable methods. The uncertainty of certified value was evaluated from each methods to be used.

Example 3. GBW08403 PAHs in coal fly ash.

The material was collected from a boiler fired with coarse coal. PAHs were extracted in Soxhlet extractor with benzene for 24h and the cycle time about 20 min. Capillary gas chromatography with flame ionization detection (GC/FID) and high-performance liquid chromatography with fluorescence detection (LC/FLD) were used in the certification of PAHs. The quantification was based on external standards. The certified results are shown in table 12.

Table 11 *CRMs of Tobacco*

Code number	GBW08514		GBW08515		Methods to be used in certification
Element or component	Certified value	Uncertainty	Certified value	Uncertainty	
Ca (%)	3.00	0.06	4.29	0.14	AAS,ICP,IC
Mg	0.51	0.02	0.74	0.04	AAS,ICP,IC
K	2.39	0.08	3.31	0.10	AAS,ICP,IC
Element or component	Certified value	Uncertainty	Certified value	Uncertainty	
N	1.72	0.04	3.58	0.10	IC, kaj-N
Cl	0.77	0.02	0.98	0.05	IC, VIS-SP
P	0.229	0.012	0.261	0.019	IC, VIS-SP
Mn(g/g)	94.3	4.2	23.4	13	AAS,ICP
Cu	16.2	1.8	17.8	1.7	AAS,ICP
Zn	28.7	2.2	36.7	3.1	AAS,ICP
Fe	95.9	4.6	96.3	50	AAS,ICP
B	21.8	1.8	41.8	2.0	ICP,COL
Nicotine	1.73	0.04	3.84	0.13	GR,UV-SP
Total Sugar	18.8	0.6	1.8	0.3	VOL,COL
Reduced Sugar	17.3	0.6	/	/	VOL,COL

Table 12 *Certified Value of GBW08403*

Component	Certified value (g/g)
Phenanthrene	7.12.6
Anthracene	2.00.8
Fluoranthene	7.41.9
Pyrene	72
Benzo[a]pyrene	1.30.3

The certified values were the average of the results obtained from two independent methods, the uncertainties were evaluated based on type A and type B standard uncertainties and expressed by 2 u_c.

4.4 Gas mixtures

There are more than hundred gas mixtures including inorganic and organic component listed in GBW RMs catalogue. They serves the public in environmental monitoring, quality control of petrochemical and chemical factories, calibrating analytical instruments of scientific and technical research institutions, etc.

Three techniques are used in preparing gas mixtures, the first one is gravimetry. This is a primary method. When it is used to prepare gas mixtures, the key point is to know the impurities in sample gas and diluting gas. It is true that to use a stable, reproducible and high-precision balance is important, but to study and determine the impurities in sample and diluting gases is more important, because the impurities will effect the accuracy of certified value and stability of the gas mixture. A 30Kg/1mg balance and first class weights were used to produce PRM grade gas mixtures (See table 13,14). Some methods for analyzing impurities of O_2, H_2O,CO, CO_2, NO, CH_4, Ar, such as API-MS, GC, UV, IR and chemiluminescence (CLU) etc. were studied. Usually the gas mixtures must undergo experiment test after preparation. So GC, IR, UV and CLU are also used in inspections. The second technique is volumetry or pressure method for CRM and WRM, the certified value is obtained through analysis of different methods. The third

Table 13 *PRM of Gas Mixture (multi-component)*

Code number	Component	Certified value (mol/mol)		Relative uncertainty %
GBW08131	C_2H_6,CH_4,C_3H_8,C_2H_4,i-C_4H_4 in N_2	C_2H_6	1000	
		CH_4	1000	
		C_3H_8	100	1.5
		C_2H_4	10	
		i-C_4H_{10}	10	
GBW08132	CH_4,C_2H_6, C_3H_8 C_3H_6 in N_2	CH_4	500	
		C_2H_6	500	1.5
		C_3H_8	10	
		C_3H_6	10	

Table 14 *PRM of Gas Mixture (single component)*

Code number	Component	Certified value (μmol/mol)	Relative uncertainty (%)
GBW08101	CH_4/N_2	10	
GBW08102		50	
GBW08103		100	1
GBW08104		500	
GBW08105		1000	
GBW08106	CO/N_2	10	
GBW08107		50	
GBW08108		100	1
GBW08109		500	
GBW08110		1000	

Code number	Component	Certified value (μmol/mol)	Relative uncertainty (%)
GBW08111	CO_2/N_2	10	
GBW08112		50	
GBW08113		100	1
GBW08114		500	
GBW08115		100	
GBW08116	NO/N_2	50-2000	1
GBW08117	O_2/N_2	21% (mol)	0.1
GBW08119	CH_4/Air	1-100	1
GBW08120	CO/Air	5-50	1
GBW08121	CH_4/Ar	10	1.3

Table 15 *PRMs of Permeation Device*

Code number	Component	Permeation rate μg/min	Uncertainty %	Used temperature ^oC
GBW08201	SO_2	0.37-1.4	1	
GBW08202	NO_2	0.6-2.0	1	
GBW08203	H_2S	0.1-1.0	2	25
GBW08204	NH_3	0.1-1.0	2	
GBW08205	Cl_2	0.2-2.0	2	
GBW(E)080049	SO_2	0.1-1.5	2	50
GBW(E)080046	SO_2	0.1-1.0	1	25,30,35
GBW(E)080050	NO_2	0.5-2	2.5	50
GBW(E)080047	NO_2	0.1-1.0	1	25,30,35

one is permeation device for low concentration and special gases (Table 15). The permeation rate was measured using high-precision balance.

The uncertainty of certified value consists of type A and type B standard uncertainty and is expressed by $3\,u_c$.

5 PROPER USE OF REFERENCE MATERIALS

5.1 The basic principle of choosing RMs

When users choose RMs according to their purpose, they must confirm the suitability of RMs with respect to certified value and uncertainty, certifying methods, date of certification, statement of intended use, expiration date, minimum sampling size, storing conditions and special instruction for correct use given in the certificate.

The followings are the basic principle to be considered for choosing RMs.

5.1.1 Matrix of RM. The matrix effect is often found in analytical measurement, especially in clinic and biochemical measurements. The first principle for choosing RM is the matrix of RM should be same or similar to that of sample to be measured, for instance, determining some elements in human serum, if a aqueous solution standard was used in calibration, wrong results may be obtained because of the effect of matrix. For such measurement a human serum RM should be better than the aqueous solution for making calibration curve.

5.1.2 Form of RM. RMs may be solid, liquid or gas, the solid RMs are usually in different forms. To select the form of solid RM based on the measuring method is also one of the basic principles. For instance, the X-ray fluorescent spectrometry usually uses RM in piece, when selecting RMs for this measurement, the surface state should be paid attention to, it is better to have it as same as that of the sample.

5.1.3 Certified value and uncertainty of RM. The certified value and uncertainty of RM should meet the demands of use and comply with economic and reasonable principles. The higher the accuracy of the property value is, the better the result could be got. But seeking higher accuracy excessively means wasting time and money. Metrologists usually use one third (1/3) principle when choosing metrological standard, i.e. the uncertainty of metrological standard should not exceed 1/3 of uncertainty of practical measurement. This may take as a reference of choosing RMs.

5.1.4 Homogeneity and stability of RM. For homogeneity of RM the essential point to be considered is whether the minimum sampling size pointed on its certificate can match the requirement. If the practical sampling size is less than the minimum sampling size of RM, it may cause heterogeneous error. When considering the stability of RM, the valid time must suit the practical demand, especially for long-term quality control.

5.2 The requirements for using RM

Before using a RM, users are required to read the certificate of RM carefully, to store, handle and use RMs following the instruction. Based on the purpose of use it is required to design a suitable experiment procedure for correctly estimating and concluding the measuring results; and to check if the analytical method, operation procedure and measuring conditions are in normal state (i.e. under statistical control). When all these are under statistical control, the measuring results can be given based on measuring data and uncertainty of certified value of RM used.

5.3 Some techniques of using RMs

RMs are used in different fields, serve for different purpose. In general, RMs may be used in 4 ways: calibration standard, working standard, quality assurance and arbitration.

5.3.1 RMs used in calibration. There are two cases to use RM for calibration: one is calibration of measuring instrument, another is calibration of analytical procedure. The calibration of instrument mainly include precision, sensitivity, responsive curve, detectable limit and stability etc. when calibrating an instrument, a set of concentrations of RMs covering the measuring range of the instrument should be used, under the normal operating conditions of the instrument, measure them from low to high concentration. When calibrating the analytical procedure for setting up the quantitative relation between signal and concentration of real sample, a series of RMs with similar matrix to that of the sample should be selected and measure them with same method under same condition of sample measurement. The linear regression may be used for data processing. All original data, estimated value $\hat{a}, \hat{b}$ of a and b in linear equation $y = a + bx$, the coefficient of correlation and formulas for calculating standard deviation are summarized in table 16.

Table 16 *Data and Equations for Linear Regression Calculation*

No. of measurement	Certified value of RM	Response value of instrument	$x_i - \bar{x}$	$(x_i - \bar{x})^2$	$y_i - \bar{y}$	$(y_i - \bar{y})^2$	$(x_i - \bar{x})(y_i - \bar{y})$
1	x_1	y_1	$x_1 - \bar{x}$	$(x_1 - \bar{x})^2$	$y_1 - \bar{y}$	$(y_1 - \bar{y})^2$	$(x_1 - \bar{x})(y_1 - \bar{y})$
2	x_2	y_2	$x_2 - \bar{x}$	$(x_2 - \bar{x})^2$	$y_2 - \bar{y}$	$(y_2 - \bar{y})^2$	$(x_2 - \bar{x})(y_2 - \bar{y})$
3	x_3	y_3	$x_3 - \bar{x}$	$(x_3 - \bar{x})^2$	$y_3 - \bar{y}$	$(y_3 - \bar{y})^2$	$(x_3 - \bar{x})(y_3 - \bar{y})$
.	.	.	.	.	.	.	.
.	.	.	.	.	.	.	.
m	x_m	y_m	$x_m - \bar{x}$	$(x_m - \bar{x})^2$	$y_m - \bar{y}$	$(y_m - \bar{y})^2$	$(x_m - \bar{x})(y_m - \bar{y})$
	$\sum x_i$	$\sum y_i$	$\sum(x_i - \bar{x})$	$\sum(x_i - \bar{x})^2$	$\sum(y_i - \bar{y})$	$\sum(y_i - \bar{y})^2$	$\sum(x_i - \bar{x})(y_i - \bar{y})$

$$\bar{x} = \frac{1}{m}\sum x_i, \quad \bar{y} = \frac{1}{m}\sum y_i,$$

$$\hat{b} = \frac{\sum(x_i - \bar{x})(y_i - \bar{y})}{\sum(x_i - \bar{x})^2}$$

$$\hat{a} = y - \hat{b}\bar{x}$$

$$S_f = \left[\frac{\sum(y_i - \hat{y}_i)^2}{m-2}\right]^{1/2}$$

$$\hat{y}_i = \hat{a} + \hat{b}x_i$$

$$\rho = \frac{\sum(x_i - \bar{x})(y_i - \bar{y})}{[\sum(x_i - \bar{x})^2(y_i - \bar{y})^2]^{1/2}}$$

$$S_0 = S_f\left[1 + \frac{1}{m} + \frac{(x_0 - \bar{x})^2}{\sum(x_i - \bar{x})^2}\right]^{1/2}$$

substitute $\hat{x}_0 = \dfrac{y_0 - \hat{a}}{\hat{b}}$ for x_0

$$S_{x'} = S_f\left[\frac{1}{n} + \frac{1}{m} + \frac{(x' - \bar{x})^2}{\sum(x_i - \bar{x})^2}\right]^{1/2}$$

substitute $\hat{x}' = \dfrac{\bar{y}' - \hat{a}}{\hat{b}}$ for x',

$$\bar{y}' = \frac{1}{n}\sum y_i$$

The analysis of linear regression is based on three assumption: y is a chance variable conformed to normal distribution, x is a fixed value which is very closed to true value; y is a dependent variable of x. If using the equation $\hat{y}_i = \hat{a} + \hat{b}x_i$ to compute the property value x of sample, the $\hat{b}$ and remained standard deviation S_f must not change significantly, therefore, it requires that the matrix of calibration standard should match the real sample as much as possible; the measuring method, operating procedure and measuring conditions for making calibration curve should be same as that of sample measurement.

In addition, there are no linear relations between x and y some times, but a linear relation can be produced through functional transformation. For example, The relation between darkness difference of spectrometric line recorded on a sensitive plate, $(\frac{\Delta s}{r})$ and the concentration c of measured element in atomic emission spectrometric analysis is not linear but logarithmic, $\frac{\Delta s}{r} = \log a + b \log c$ let $y = \frac{\Delta s}{r}$, $\log c = x$, $\log a = a$, then the above equation is changed into $y = a + bx$, this is linear equation. Besides the logarithmic transformation, hyperbola, parabola, exponential and power function, etc. could be changed into linear relation through certain transformations.

Let's take verification of an instrument and calibration of a analytical procedure as the example to illustrate the use of RMs.

1) Verification of an analytical instrument

When verifying the linearity , sensitivity, precision and stability of an instrument, the first thing to do is to choose suitable RMs of 5 or more concentrations in gradient according to the detectable range of the instrument, which may be pointed on its instruction manual; the second is to measure the RMs under normal operating conditions. The arrangement of experiment, data analysis ,conclusion of verification, etc. can be carried out following the process in Table 17.

Table 17 *Process of Verification or Calibration*

Experiment		Data analysis	Conclusion
Certified value of RM	Obtained value		
x_1	$y_{11}, y_{12} \cdots y_{17}$ ①		S_y is precision of
x_2	y_2		lower limit of
x_3	y_3	$S_y = [\dfrac{\Sigma(y_{ij} - \bar{y})^2}{7-1}]^{1/2}$,	quantitative analysis; sensitivity is equal
•	•		to $\hat{b}/s_f$;
•	•	② Calculate $\hat{a}, \hat{b}, s_f, p$ using the	
•	•	equations in table 16.	linearity: $y_i = \hat{a} + \hat{b}x_i$
x_m	y_m	③ Check significance with	range of linearity:
		respect to relative coefficient. If	$x_1 - x_m$
		p>0.959, then the linear relation	
		exits between x and y with 99%	
		certainty.	

Experiment			*Conclusion*
		Data analysis	

Measuring again several hours later:

x_1	y_1	① analyze and calculate the data
x_2	y_2	obtained refer to table 16,
x_3	y_3	$\hat{a}, \hat{b}, S_f$
x_4	y_4	② calculate standard deviation
•	•	of $\hat{a}, \hat{b}$, , with equations of :
•	•	
x_m	y_m	

$$S_a = \left[\frac{S_f^2 \sum X_i^2}{m \sum (x_i - \overline{x})^2} \right]^{1/2}$$

$$S_b = \left[\frac{S_f^2}{\sum (x_i - \overline{x})^2} \right]^{1/2}$$

③ test the two regression lines obtained, using f test to check identity of remained standard deviation, using t test to check identity of $\hat{a}$ and $\hat{b}$.

Stability: If both f and t test were passed, that means the two regression lines obtained by two times measurements are identity, the stability of the instrument is good.

2) Calibration of an analytical procedure

When RMs are used for calibration of a analytical procedure the matrix of RM should match the sample to be measured, the range of certified values should cover the property values of the sample, the calibration curve should be made with a series concentrations of RM under the same measuring conditions of the sample. The data processing may followed the form and equations in table 16. If the linear relation is got, then the sample can be measured. Substituting the data obtained into $y_0 = \hat{a} + \hat{b}x_0$, the concentration X_0 of sample can be calculated. The measurement result is expressed by $X_0 \pm t_{0.95}S_{xo}$, which is 95% confidence interval. If the sample is measured n times, the result will be $X' \pm t_{0.95}S_{x'o}$ in 95% confidence interval.

5.3.2   *RMs used as working standard.* When RMs are used as working standards, it is just like weighing substance using weights. Users should choose a RM whose matrix and certified value close to that of sample to be measured. With same instrument, procedure and working conditions, measure RM and sample alternatively, or one RM every 2 or 3 samples, then calculate the property value of sample using the following equation:

$$X_{samp} = \frac{Y_{samp}}{Y_{std}} X_{std.}$$

The uncertainty of the result is

$$u_{samp} = (u_{std.}^2 + \frac{S^2}{n})^{1/2}$$

or $u_{samp} = (u_{std.}^2 + S^2)^{1/2}$

The final result is $X_{samp} \pm u_{samp.}$

In practice, it is hard to choose a RM with the matrix and property value very close to that of sample. In this case people often use the method of adding standard to the sample to get accurate results. The outline of this method is: take certain volume (V_x) of test solution for sample, measure it and get a result Y_x, then add a certain volume (V_s) of standard solution into test solution, measure it again and get result Y_{x+s}. Using C_x and C_s to present the concentrations of sample and standard solution respectively then C_x can be calculated:

$$C_x = \frac{V_s C_s Y_x}{V_x \left(Y_{x+s} - Y_x\right)}$$

The key point of this method is the concentration of the standard solution should be higher, and the added volume must be as small as possible to ensure the negligible change of the concentration and matrix of test solution of sample.

5.3.3 RMs used in program of quality assurance. The major purpose of quality assurance in analytical measurement is to make people to believe the reliability of measuring results. The RMs with accurate value of a quantity can provide fair, scientific and authoritative bases, so that using RMs for quality assurance is the most suitable choice. There are 3 techniques in this case:

1) The person in responsible position or entrust side may choose a RM which is similar to the sample to be measured and give to operators as a secret code sample. After measuring compare the result $\overline{X} \pm t_{0.95} S$ with the certified value of RM, A $\pm u_c$ if $|\overline{X} - A| \leq 2\left(u_c^2 + S^2\right)^2$ then the analytical results are reliable, otherwise there may be bias or mistakes in measuring process.

2) When the sample is valuable or rare, a parallel measurement of RM and sample should be made, then judge the reliability based on the measuring results.

3) When a laboratory carries out a long term routine analysis which is significance for economy and society, the one who is in responsible position should take a appropriate RM to plot a chart of long-term quality control in order to help finding out and handling problems occurred during the analytical measurements. The procedure in detail is that the organizer gives a RM in secret code together with samples to operators, collect results after measuring, take the certified value of RM as central line, two times of standard deviations of reproducibility, $2S_R$, as up and low control limit, with date or order of measurement as abscissa, draw a diagram of accuracy control.

If all results obtained fall into the range of control limit randomly, then the measuring results are correct, the uncertainty of results is expressed with $2S_R$. If the results fall into the range but tend to up or down, then it needs to check the measuring process and condition,find out reason and resolve the problem existed timely . Another case which should be paid more attention to is that several results got in succession are same. This is not accord with statistical regulation, it may be caused by operators.

5.3.4 RMs used in metrological arbitration. The quality disputes in international and domestic trade happen often. So metrological arbitration is required for the judgment. If the arbitrator could use an appropriate RM, then the arbitration can be more impartial. When a RM is used in this work, the arbitrator can hand over it in secret code both side of disputations to measure, adjudge the reliability of measurements based on whether the submitted data consist with the certified value of RM. Arbitration using RM is more economic, authoritative, directive and objective than inviting the third side to do arbitrary analysis.

References

1. BIPM, ISO, etc., International Vocabulary of Basic and General Terms In Metrology, 1993, Geneva.
2. W. Richter, Report of the comite consulttif pour la Quantite de Matiere (st Meeting), 1995, Paris.
3. ISO Guide 30, Terms and definitions used in connection with reference materials,1992, Geneva.
4. CITAC, Guide 1, Guide to Quality in Analytical Chemistry, 1995, UK.
5. BIPM, ISO, etc., Guide to the Expression of Uncertainty in Measurement, 1993, Geneva.
6. IUPAC/ISO/AOAC, Harmonised Guidelines for Internal Quality Control in Analytical Chemical Laboratories,1994.
7. M. Thompson, R. Wood , Pure. Appl. Chem., 1993.**65**, 2123.
8. ISO/IEC, Guide 25, General requirements for the competence of calibration and testing laboratories, 1990, Geneva.
9. ISO 10012, International standards for quality assurance requirements for measuring equipment, 1992, Geneva.
10. ISO Guide 35, Certification of reference materials, General and statistical principles,1989,Geneva .
11. ISO Guide 33, Uses of certified reference materials,1989, Geneva.
12. ISO Guide 34, Quality system guidelines for the production of reference materials,1996, Geneva.
13. Pan XiuRong, VAM Bulletin, 1995, **12**,18.
14. Pan XiuRong, Accred. Qual. Assur., 1996, **1**,181.
15. Pan XiuRong, Metrologia, 1997, **34**,35.

How to Use Matrix Certified Reference Materials? Examples of Materials Produced by IRMM's Reference Materials Unit

J. Pauwels

EUROPEAN COMMISSION – D.G. JOINT RESEARCH CENTRE, INSTITUTE FOR REFERENCE MATERIALS AND MEASUREMENTS (IRMM). REFERENCE MATERIALS UNIT, B-2440 GEEL, BELGIUM

1 INTRODUCTION

The Institute for Reference Materials and Measurements (IRMM) is the standards institute of the Joint Research Centre of the European Commission. It produces and certifies both nuclear and non-nuclear certified reference materials (CRM's) of chemical and isotopic nature, and is responsible for the distribution, control and renewal of BCR CRM's, which are to the largest extent produced under the auspices of the Standards, Measurements and Testing (SMT) Programme [X] of the European Commission.

Isotopic CRM's, designated as IRM's, are in principle produced in-house under the responsibility of the "Isotope Measurements"-unit: isotope ratio CRM's via absolute mass spectrometric measurements calibrated using synthetic isotope mixtures; spike CRM's for use in isotope dilution mass spectrometry (IDMS) via inverse IDMS using high purity chemicals. CRM's certified for chemical composition (including CRM's for reactor neutron dosimetry) are produced by the "Reference Materials"- and the "Analytical Chemistry"-units. They are, in general, certified on the basis of laboratory intercomparisons involving a variety of methods based on different physical and/or chemical principles.

Certified values are expected to be correct - with a probability of 95 % - within the stated uncertainty intervals, but in practice the quoted uncertainties may have quite different meanings and may be based on quite different principles. It must be admitted that neither the differences nor their practical usefulness for the user are always evident. BCR CRM's have however the distinct advantage that they are, in general, distributed not only with a certificate, but also with a detailed certification report providing all available analytical and statistical information. In many cases, the certification report also contains specific information on how to use the certified data e.g. for method assessment.

ISO Guide 33 on the "Use of reference materials"[1] states that " CRM's must be used on a regular basis to ensure reliable measurements". In reality, the expression "to ensure reliable measurements" covers various meanings, such as "to transfer information on

---

[X] Previously: MAT-programme (1992-94), BCR-programme (1988-91), Bureau Communautaire de Référence (<1988)

property values", "to assure traceability to unit scales or to standards", or "to assess precision and/or trueness of measurement processes". Although these different applications of CRM's may require different information from the producer, CRM's, mostly, just carry a certified value and an uncertainty value, which is generally *said to be* a 95 % confidence interval or something similar, and which relevance - as was stated by Jorhem[2] at the BERM-7 Symposium in Antwerp, Belgium - is not always evident for the user. The reasons for it are multiple and certainly not the sole responsibility of the certifying agency, which is generally following international guidelines, such as the various guides prepared under the auspices of ISO-REMCO, which not only concentrate on a certified value "accompanied by an uncertainty at a stated level of confidence"[3], but still ignore the international acceptance of the Guide to the expression of uncertainty in measurement (GUM)[4]. The responsibility for the ignorance by most users of the ISO Guides related to CRM's can, however, not be put on the shoulders of the producers as Jorhem does: first ISO is an internationally known and recognised organisation with responsible delegates in most countries world-wide, second e.g. BCR explicitly refers in many of its environmental CRM - reports with a short explanation to ISO Guide 33; finally IRMM took the initiative to include the ISO - REMCO brochure "The role of reference materials in achieving quality in analytical chemistry"[5], with references to all existing ISO Guides in the field, in its 1998 catalogue of BCR Reference Materials[6] which is distributed world-wide to more than 10 000 potential users of CRM's.

2 SOME EXAMPLES OF CRM's PRODUCED BY IRMM's RM - UNIT

2.1 BCR CRM 278R: Trace Elements in Mussel Tissue[7]

CRM 278R is an environmental BCR reference material, certified for trace elements in a natural biological matrix. It is the successor of BCR CRM 278, which was certified in 1988, and is now exhausted. It is intended for method validation and quality control of marine environmental monitoring, in which fish, mussels and aquatic plants are considered to be indicators for the contamination of the environment they live in. The CRM was prepared from mussels collected in the Dutch Waddensea, and is available in the form of a stable, homogeneous dry powder prepared by freeze drying, ball milling, sieving and bottling in 8 gram amounts in brown glass vials under inert atmosphere. To increase the long-term stability of the material under normal laboratory conditions, its residual moisture content was reduced to 1.3 ± 0.2 %. The material was controlled for homogeneity by instrumental neutron activation analysis, hydride generation atomic absorption spectrometry and/or inductively coupled plasma - mass spectrometry. Stability measurements were carried out at different temperatures (40 °C, 18 °C and 4 °C) relative to measurements carried out at a reference temperature of $- 20$ °C at which the stability of the material is supposed to be guaranteed. Both in the homogeneity and stability measurements, the observed variations were comparable to the expected method variations (C.V. < 3 %); subsequently, they were left out of further consideration, as is normal praxis for BCR CRM's certified in the SMT - programme.

BCR CRM 278R was certified for nine trace elements (Al, Cd, Cr, Cu, Hg, Mn, Pb, Se and Zn) on the basis of laboratory intercomparisons carried out according to the Guidelines for the production and certification of BCR reference materials[8]. More particularly, the laboratories participating in the certification used at least the following quality control steps to verify if their methods were applied correctly:

- special consideration was given to calibration of equipment and the verification of the composition and traceability of calibrants;
- where digestion techniques were used, the participants either took all precautions to ascertain a total digestion (treatment with HF) or verified that the residue after digestion did not contain the element to be determined;
- the determinations were only performed when the method was under statistical control;
- whenever possible, the performance of the method was verified by analysing available CRM's (in most cases BCR CRM 278, which is very similar to the CRM 278R) or other well characterised materials of similar matrix- and trace element composition. After technical scrutiny of the results of each individual laboratory, a statistical treatment was applied (Table 1) and certified values were calculated (Table 2). The quoted uncertainties are the half-widths of the 95 % confidence intervals, calculated on the basis of the mean value of each laboratory.

It must, however, be clear that the quoted confidence interval (which correctly describes the statistical uncertainty of the certified parameter!) cannot be used as such neither for calibration nor for validation according to ISO Guide 33:

- *Calibration* (in so far matrix CRM's should be used for this purpose) requires a standard deviation (or a combined standard uncertainty as now defined in the GUM) for inclusion in the analysis uncertainty budget[9]. This value is however available in the certification report, and can, anyhow, easily be calculated by dividing the certified value by the student factor t (for ν, the number of degrees of freedom, equal to the number of laboratories minus 1).

Table 1 *Summary of statistical data for BCR CRM 278R, mussel tissue (mass fractions in mg/kg)*

	As	Cd	Cr	Cu	Hg	Mn	Pb	Se	Zn
Data sets	10	9	8	9	4	9	10	8	11
Accepted replicates	59	54	44	54	24	55	60	49	67
Outlying variances (a)	no	no	no	no	no	no	no	no	no
Homogeneity of variances (b)	yes	no	yes	yes	yes	yes	yes	yes	no
Normal distribution of sets (c)	yes	yes	yes	yes	yes	yes	yes	yes	yes
Outlying means (d)	no	no	no	no	no	no	no	no	no
Mean of means (e)	6.07	0.348	0.78	9.45	0.196	7.69	2.00	1.84	83.1
s - within (e)	0.21	0.017	0.043	0.31	0.009	0.26	.055	0.085	1.39
s - between (e)	0.16	0.005	0.066	0.11	0.004	0.27	.038	0.105	2.38
95 % C.I. of mean of means (e)	0.13	0.007	0.06	0.13	0.009	0.22	0.04	0.10	1.7

(a) Cochran test   *(c) Kolmogorov-Lilliefors test*   *(e) results in mg/kg*
(b) Bartlett test   *(d) Nalimov t-test*

Table 2 *Certified values of trace elements in BCR CRM 278R, mussel tissue*

Certified element	Mass fraction (mg/kg)
As	6.07 ± 0.13
Cd	0.348 ± 0.007
Cr	0.78 ± 0.06
Cu	9.45 ± 0.13
Hg	0.196 ± 0.009
Mn	7.69 ± 0.23
Pb	2.00 ± 0.04
Se	1.84 ± 0.10
Zn	83.1 ± 1.7

- If no "requirement -, legal -, accreditation -, user - or experience limits" are imposed on the analysis laboratory, the **assessment of the precision of a measurement process** according to ISO Guide 33 requires a within-laboratory standard deviation and is obtained by computing a χ^2_c value equal to:

$$\chi^2_c = (s_w / \sigma_{wo})^2 \tag{1}$$

with:

  s_w =   the within-laboratory standard deviation under repeatability conditions obtained in the assessment experiment

  σ_{wo} =   the required value of the within-laboratory standard deviation (possibly to be taken from the certification if carried out by laboratory intercomparison)

and comparing this value χ^2_c with a χ^2_{table} - value obtained as:

$$\chi^2_{table} = \chi^2_{(n-1)\,;\,0.95} / n -1 \tag{2}$$

If $\chi^2_c \geq \chi^2_{table}$ there is no evidence that the measurement process is not as precise as required (or as was achieved by the laboratories in the certification exercise); if, on the contrary, $\chi^2_c < \chi^2_{table}$ there is evidence that the measurement process is not as precise as required.

- If the assessment experiment is performed by only one laboratory, and thus the between-laboratory fluctuation σ_{Lm} cannot be determined directly, the **assessment of the trueness of a measurement process** according to ISO Guide 33 requires the between-laboratory standard deviation σ_L of the certification. For the assessment of trueness, the following general condition is used as the criterion for acceptance:

$$- a_2 - 2\,\sigma_D \leq \bar{x} - \mu \leq a_1 + 2\,\sigma_D \tag{3}$$

with:

x̄   = average value found by the laboratory

μ = certified value of the CRM

a_1 and a_2 = adjustment values chosen in advance by the experimenter according to economic or technical limitation or stipulation;

σ_D = the standard deviation associated with the measurement process.

σ_D is composed of two parts:
- the within-laboratory or short-term fluctuation estimated by s_w;
- the between-laboratories fluctuation with standard deviation σ_{Lm}, estimated by the certification between-laboratory standard deviation σ_L;

and is given by:

$$\sigma^2_D = \sigma^2_{Lm} + s^2_w / n \qquad [4]$$

In the case of BCR CRM's, this information is, in general, available to the user, if not in the certificate, then in the accompanying certification report. Moreover, for environmental CRM's a short specific chapter generally focuses the attention of the user on this point, thereby referring explicitly to ISO Guide 33.

2.2 BCR CRM 462: Butyltins in Coastal Sediment[10]

BCR CRM 462 is an environmental matrix reference material, which was certified in 1994 under the SMT - programme, but had to be withdrawn from sales in 1996 due to stability problems at the long-term storage temperature of 4 °C. In 1996 the batch was transferred to - 20 °C, and recertified under the responsibility of IRMM. CRM 462 is intended for quality control purposes and for the validation of analysis methods used in the frame of legislative controls related to the use of TBT - based antifouling paints. The reference material was prepared from coastal sediment collected in the southern part of the Arcachon Bay in France, and is available in the form of homogenised dry powder prepared by air drying at 55 °C, coarse sieving through a 1 mm mesh size sieve, jet-milling with classification, sterilisation at 120 °C, homogenisation by turbula mixing and bottling in 25 g units in brown glass bottles with polyethylene inserts. To increase the long-term stability under normal laboratory conditions, the residual moisture content was reduced to 2.7 ± 0.4 %. The material was controlled for homogeneity by GC-MS and found to be sufficiently homogeneous for sample intakes of 0.5 g or larger. The stability at the new storage temperature of - 20 °C was controlled, using the same method, during a period of 12 months relative to samples stored at a reference temperature of - 70 °C and will be further monitored at regular (yearly) intervals.

The material was certified for tri- and dibutyltin (monobutyltin could not be certified due to a lack of agreement between the results of the various laboratories) on the basis of laboratory intercomparisons carried out in respect of the Guidelines for the production and certification of BCR reference materials[8]. More particularly, the laboratories participating in the certification used at least the following quality control steps to verify if their methods were applied correctly:

- special consideration was given to the calibration of the equipment and the verification of the composition and traceability of the calibrants: therefore, a set of ultrapure calibrants, synthesised and purified in large quantities by an expert laboratory were distributed to the certifying laboratories for quality control;

- all steps involved in the analytical process, i.e. sampling, extraction, clean up, derivatisation, separation, detection, calibration were discussed in great detail to avoid possible sources of error and find appropriate measures to eliminate them;
- the determinations were performed only when the method was under statistical control;
- whenever possible, the performance of the method was verified by analysing available CRM's or other well characterised materials of similar matrix- and trace element composition.

After technical scrutiny of the results of each individual laboratory, a statistical treatment was applied (Table 3) and certified values were calculated.

However, as a consequence of the insufficient repeatability of the homogeneity measurements, the absence of significant between-units variation could not be positively demonstrated. Therefore, a between-units standard uncertainty was estimated according to Pauwels et al.[11] and included in the CRM uncertainty (U), obtained as:

$$U = 2 \cdot u_c = 2 \cdot (u_1^2/N + u_2^2)^{1/2} \tag{5}$$

with:

u_1 = standard deviation of the laboratory mean values
u_2 = estimated between-units standard uncertainty
N = number of participating laboratories

The resulting certified values and uncertainties are summarised in Table 4.

As in the previous example, the quoted uncertainty (here an expanded uncertainty according to the ISO - BIPM Guide[4] with coverage factor k = 2) cannot directly be used as

Table 3 *Summary of statistical data for the recertification of BCR CRM 462 (mass fractions in μg/kg)*

	TBT	DBT
Data sets	5	7
Accepted replicates	29	40
Outlying variances (a)	no	no
Homogeneity of variances (b)	yes	no
Normal distribution of sets (c)	yes	yes
Outlying means (d)	no	no
Mean of means (e)	54.5	68.4
s - within (e)	6.3	9.7
s - between (e)	3.0	10.1
95 % C.I. of mean of means (e)	5.1	9.5

(a) Cochran test   *(c) Kolmogorov-Lilliefors test*   *(e) results in mg/kg*
(b) Bartlett test   *(d) Nalimov t-test*

Table 4   *Newly recertified values of butyltins in BCR CRM 462, coastal sediment*

Certified compound	Mass fraction (µg/kg)
TBT	54 ± 15
DBT	68 ± 12

such, but:

- For **Calibration** purposes a combined standard uncertainty for inclusion in the analysis uncertainty budget[11] is readily available by dividing the certified uncertainty by the coverage factor used (i.e. 2);
- For the **assessment of precision** according to equations [1] and [2] or the **assessment of trueness** according to equations [3] and [4], the values of σ_{wo}, the within-laboratory standard deviation, or of σ_L, the between-laboratory standard deviation, can be taken from the certification measurements described in the certification report. Also in this case, all information is available in the accompanying certification report, which contains a short specific chapter on how to use CRM 462, thereby referring to ISO Guide 33.

2.3 VDA CRM's 001-004: Cadmium in Polyethylene[12]

The VDA CRM's 001-004 constitute a set of polyethylene reference materials doped with four different mass fractions of cadmium. They were certified by IRMM on behalf of the Verband der Automobilindustrie e.V. (VDA), Frankfurt, Germany, and are intended for quality assurance purposes in the frame of legal restrictions of cadmium contents in different parts of automobiles. Four different batches of doped polyethylene were prepared for this project by BASF Farben und Lacke, Germany. The material was made available in the form of small wire cuts of approximately 10 mg each and was subsequently homogenised by turbula mixing and bottled in 30 g units at IRMM. The homogeneity of the reference materials was verified using solid sampling atomic absorption spectrometry[13]. On the basis of these measurements, minimum sample sizes (M) for which the certified values are valid could be determined for VDA 001 - 003, according to the method described by J. Pauwels and C. Vandecasteele[14]. The results are summarised in Table 5.

The reference materials were certified on the basis of thermal ionisation isotope dilution mass spectrometry, according to ISO Guide 35: 1989 (E) § 7 (Certification by a definitive method)[15], using two ^{111}Cd spike solutions characterised by reverse IDMS using 99.999 % pure cadmium metal. A summary of the IDMS results is given in Table 6. The certified values are obtained in a way guaranteeing full traceability to the SI-system of units, and are corrected for all known sources of systematic error. The quoted uncertainties account for all identified sources of uncertainty. To the statistical uncertainty given by the repeatability of the Cd determinations, an uncertainty of systematic nature, taking into account all factors having a systematic influence on the results, was added. This uncertainty contribution of systematic nature amounts to 0.70 - 0.75 % (estimated equivalent of one standard deviation). The total uncertainty was computed by adding to the statistical component (95 % confidence interval according to ISO Guide 35) an

Table 5 *Minimum representative sample mass determined by solid sampling atomic absorption spectrometry*

CRM	Min. sample M (mg)
VDA - 001	27
VDA - 002	18
VDA - 003	13
VDA - 004	*

* *the SS-ZAAS results obtained on VDA - 004 were not normally distributed and distinct heterogeneities could be observed; however, as no trends could be detected between bottles, the material remains useful provided no samples smaller than 1 gram are used*

Table 6 *Summary of the results of the IDMS measurements (mass fractions in mg/kg)*

CRM	VDA - 001	VDA - 002	VDA - 003	VDA - 004
Mean value	40.9	75.9	197.9	407
IDMS repeatability *	0.5	1.0	2.0	10
Samples analysed	6	6	6	12
Statistical uncertainty *	0.6	1.0	2.1	6
Systematic uncertainty *	0.3	0.6	1.4	3
total uncertainty **	1.2	2.1	4.8	12

*   *standard deviation*
**   *95 % confidence level*

equivalent thereof (2 s) for the contribution of systematic nature. This way of doing, chosen at the time of certifying these CRM's, is quite conservative compared to what is presently advised by the ISO Guide to the expression of uncertainty in measurement. Complementary, the CRM's were also submitted to an international measurement evaluation programme (IMEP - 2)[16] in which participated 23 laboratories using nine different methods. The resulting values of this laboratory intercomparison turned out to be very comparable to values of the IDMS certification, as can be seen in Table 7.

Just as in the two previous examples, also here the quoted uncertainties cannot be used as such:

• The combined standard uncertainty required for ***calibration*** purposes can be computed from the statistical component of uncertainty and the systematic one mentioned in Table 6. VDA CRM's 001 - 003 are, moreover, ideally suited for calibration of solid sampling spectroscopical techniques using milligram or submilligram samples. For this purpose, however, the statistical component of uncertainty (Δ) must be increased to:

$$\Delta_m = \Delta_M \cdot \sqrt{(m/m)} \qquad\qquad [5]$$

with:

m = the mass of the microsample used in solid sampling spectroscopy;

M = the minimum representative sample mass;

the systematic component of uncertainty remaining unchanged.

- In the case of the VDA reference materials *assessment of precision of a measurement process* using the equations [1] and [2] is only possible if "requirement -, legal -, accreditation -, user - or experience limits" have been fixed because no other within-laboratory standard deviation is available than the one obtained from the IDMS measurements.

- For the *assessment of precision* according to equations [1] and [2] or the *assessment of trueness* according to equations [3] and [4], neither the value of σ_{wo}, the within-laboratory standard deviation, nor that of σ_L, the between-laboratory standard deviation, are available for the same reason as above.

Table 7 *Comparison of certification by IDMS with results of IMEP-2*

CRM	Certified value	IMEP - 2 value
VDA - 001	40.9 ± 1.2	40.7 ± 1.2
VDA - 002	75.9 ± 2.1	75.1 ± 2.1
VDA - 003	197.9 ± 4.8	197.9 ± 4.9
VDA - 004	407 ± 12	408.7 ± 8.8

- For the *assessment of trueness* using the equations [3] and [4], not only adjustment values a_1 and a_2 have to be defined, but as the certification is based on a definitive method rather than on an laboratory intercomparison, also a between-laboratories standard deviation σ_{Lm} has to be chosen by the experimenter. The adjustment values are chosen according to economic or technical limitation or stipulation, the between-laboratories fluctuation can be estimated from previous experience or from literature data (in this case IMEP-2 is of course the most appropriate source of information, although in fact no precise values were published in reference[16]).

2.4 EC-NRM 501: $^{238}UO_2$ for Reactor Neutron Dosimetry[17]

EC-NRM 501, $^{238}UO_2$ for reactor neutron dosimetry, is a reference material used for reactor surveillance and neutron metrology measurements. The reference material is available in the form of 0.5 and 1.0 mm diameter sintered UO_2 spheres and was prepared by gel precipitation and subsequent calcination at Harwell Laboratory, UKAEA, UK. The CRM was characterised at IRMM. The U mass fraction was certified by isotope dilution mass spectrometry (IDMS), potentiometric titration and gravimetry (see Table 8). The isotopic composition was determined by thermal ionisation mass spectrometry, whereas the overall purity was verified by spark source mass spectrometry and α-spectrometry.

The homogeneity of the reference material was verified by isotope dilution mass spectrometry carried out on single spheres. This resulted in observed standard deviations which can be considered as "normal" for IDMS at this level of U content, which means that no inhomogeneity could be demonstrated. Nevertheless, it was preferred not to certify the material on the basis of the 95 % confidence interval, as there was neither a positive proof of homogeneity within this obtained uncertainty. Therefore, the 95 % / 95 % tolerance interval of the homogeneity study was chosen as uncertainty for the certified value. Doing so, the certified value and its uncertainty are valid for use of one single sphere. Moreover, if dosimeter sets of N spheres are used for the dosimetry experiment, the uncertainty can be reduced by a factor $N^{1/2}$. It must however be stressed that, as specified in the certification report, an increase in sample size will not allow to reduce the CRM uncertainty below the 95 % confidence interval (879.4 ± 1.5 g/kg), which would have been certified if the material had been assumed to be homogeneous. The optimum number of spheres that can be used is therefore: $N = (2.8 / 1.5)^2 = 3.5 - 4$ spheres.

Table 8   *Summary of results of determination of U (mass fractions in g/kg)*

Method	0.5 mm spheres		1.0 mm spheres			
	mean	stand. dev.	mean	stand. dev.	mean	stand. dev.
IDMS	879.0	1.8	878.5	0.6	878.7	1.3
Gravimetry	879.7	0.2	880.0	0.3	879.8	0.3
Potentiometry	879.9	0.3	879.4	0.6	879.7	0.5

CERTIFIED VALUE = (879.4 ± 2.8) g/kg

Table 9   *Composition of CRM 599, freeze-dried curd of ewes' and goats' milk*

	0 % batch	1 % batch
Ewes' milk (%)	50.6	50.0
Goats' milk (%)	49.4	49.0
Cows' milk (%)	0	1.0

2.5 BCR CRM 599: Set of Two Curd Reference Materials for the Detection of Cows' Milk Casein in Cheeses from Ewes' and Goats' Milk[18]

BCR CRM 599 constitutes a set of two reference materials, to be used as "positive" and "negative" for the detection of cows' milk casein in cheese from ewes' milk, goats' milk and mixtures of ewes' and goats' milk according to Commission Regulation (EC) N° 1081/96. The reference materials consist of curd, prepared at the Institut Technique des Produits Laitiers Caprins, Surgères, France, starting from gravimetrically prepared mixtures of ewes', goats' and cows' milk, in the proportions mentioned in Table 9. The curd was subsequently freeze dried, ground under liquid nitrogen cooling and bottled in 15 g units in brown glass vials filled with argon at IRMM. The residual moisture content is 1.2 - 1.5 %. Homogeneity and stability were evaluated using isoelectric focusing of γ-caseins according to the EC reference method. The material is stored at - 20 °C but is sufficiently stable to be dispatched under normal postage conditions. The material has been validated for the EC reference method in a laboratory intercomparison to which participated 10 laboratories, each analysing 14 unknown and randomly distributed 0 and 1 % samples. As no false results were observed, it was concluded that the probability of obtaining a correct result is 100 %, with a 95 % confidence interval ranging from 97.8 to 100 % ($\alpha = 0.05$; n = 139).

According to Commission Regulation (EC) N° 1081/96, CRM 599 has now the status of "primary standard" for the detection of adulteration of ewes' and goats' curd by cows' milk. Unknown samples are analysed by isoelectric focusing of γ-caseins after plasminolysis in parallel with the certified 0 % and 1 % samples. The stained γ_3 and γ_2 patterns, which are the evidence of the presence of cows' milk, of the unknown sample are compared to those of the reference samples. The method is considered to operate satisfactorily if there is a clear positive signal for both bovine γ_2 and γ_3 - caseins in the 1 % reference standard but not in the 0 % reference standard. If not, optimisation of the procedure is required. A sample is judged as being positive, if both the γ_2 and γ_3 - caseins or the corresponding peak area ratios are equal or greater than the level of the 1 % reference standard.

3 BCR CRM'S AND THE "JORHEM - PARADOX"

In his critical paper on "Non-use and misinterpretation of CRM's"[2] presented at the BERM-7 Symposium in Antwerp, Belgium, Jorhem raises the paradoxical question "How results that are perfectly good for the characterization of a CRM can later be regarded as not acceptable?". As was illustrated in the above examples this question is based on a misinterpretation of the meaning of CRM uncertainties.

Taking e.g. the example of Figure 1 (As in CRM 278R), the quoted uncertainty corresponds to the half-width of the 95 % confidence interval, calculated on the basis of the mean value of each laboratory (10 laboratories, five methods, certified value: 6.07 ± 0.13 mg/kg). It can be observed that the certified interval does not overlap with the mean value of four laboratories out of 10, and does even not overlap with the 95 % confidence interval of LAB 1. In the present certification philosophy[8,15] it must, however, be understood that this is absolutely logical considering the definition of a certified reference material, which specifies that it is "a reference material, which is accompanied by a certificate, one or more of whose property values are certified by a procedure which establishes traceability to an accurate realization of the unit in which the property values are expressed, and for which each certified value is accompanied by an uncertainty at a stated level of confidence"[3], here the 95 % confidence level. According to this definition, the value which is certified is the best estimate of average arsenic mass fraction of the total batch of material, and as this value is based on laboratory mean values only, for 10 laboratories (t = 2.26, $\sqrt{10}$ = 3.16) the half-width of the 95 % confidence interval (as mentioned in the certificate) does only correspond to 71.5 % of ONE between - laboratories standard deviation. In clear, this means that, from a purely statistical point of view, only about half of the laboratory means CAN BE expected to be located in the certified interval. This is indeed the case in the example "As in CRM 278R".

It is, however, clear that this confidence interval cannot be used as such neither for calibration nor for validation purposes. Taking again, the example of Figure 1, it can,

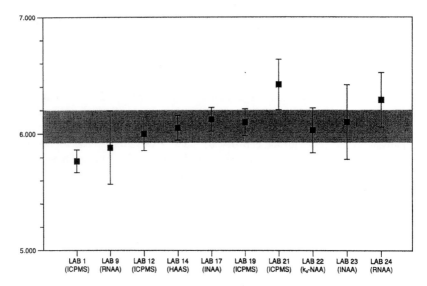

Figure 1   *Laboratory results for As in BCR CRM 278R, mussel tissue: bar-graphs represent 95 % confidence intervals (mass fractions in mg/kg)*

however, be calculated that according to ISO Guide 33:

- the laboratory with the less good standard deviation, i.e. LAB 23, has a χ^2_c-value of

$$\chi^2_c = (0.316 / 0.21)^2 = 2.26$$

to be compared to:

$$\chi^2_{table} = \chi^2_{(n-1)\,;\,0.95} / 5 = 11.07 / 5 = 2.22$$

or:

$$\chi^2_{table} = \chi^2_{(n-1)\,;\,0.99} / 5 = 15.09 / 5 = 3.02$$

indicating that there would only be a marginal evidence that the measurement process of laboratory 23 would not be as precise as required;

- the two laboratories with the most important deviation from the certified value, i.e. LAB 1 and LAB 21, have σ_D-values of:

$$\sigma_D = (\sigma^2_{Lm} + s^2_w / n)^{1/2} = (0.16^2 + 0.10^2 / 6)^{1/2} = 0.17$$

and:

$$\sigma_D = (\sigma^2_{Lm} + s^2_w / n)^{1/2} = (0.16^2 + 0.21^2 / 6)^{1/2} = 0.18$$

and, thus, 2 σ_D values of 0.33 and 0.36, which have to be compared with:

$$| \bar{x} - \mu | = | 5.77 - 6.07 | = 0.30$$

and

$$| \bar{x} - \mu | = | 6.42 - 6.07 | = 0.35$$

respectively.

This means that, as $| \bar{x} - \mu | < 2\sigma_D$ for both LAB1 and LAB 21, both results are *"perfectly acceptable"*, even taking $a_1 = a_2 = 0$.

4 WHAT SHOULD/CAN/MUST BE DONE TO HELP CRM USERS ?

Even if the "Jorhem paradox" has been clarified, as the result of LAB 1 in the above example satisfies the criterion for assessment of trueness according to ISO-Guide 33, it is still astonishing and in fact not acceptable that two results (the certified one and the laboratory one) which both claim to contain the most probable mean value of the material with a probability of 95 % do, effectively, not overlap. The main reason for this is that - in fact - neither the BCR certified value nor the laboratory value in the sense of ISO Guide 33 consider other uncertainty components than those associated with the statistical fluctuation of repetitive measurements, whereas the correct application of the ISO Guide to the expression of uncertainty in measurement would require that both for the reference material certification[19] and for the actual assessment experiment[9] ALL sources of uncertainty are considered, which, for sure, would lead to significantly larger (expanded) uncertainties, and expectedly to overlapping results. The problem is, however, that, so far, CRM's have mostly been certified and measurement results reported purely on the basis of statistics, whereas the real uncertainties of both are essentially dominated by other sources of uncertainty.

It is therefore the conviction of IRMM's RM - unit that:

- certified values of CRM's should, nowadays, no longer be based purely on statistics, but take into account *all* sources of uncertainty originating from the analytical measurements in each contributing laboratory, from possible differences between units and possible instability between certification and analysis in the users laboratory, as suggested by Pauwels et al.[19]. The CRM uncertainty should be presented in the form of an expanded combined uncertainty according to the ISO Guide to the expression of uncertainty in measurement, whereby the coverage factor should always be clearly mentioned in order to allow an easy recalculation of the combined standard uncertainty which is needed for uncertainty propagation when the CRM is used for calibration.

- measurements in "user-laboratories" should, equally, take into account *all* sources of uncertainty, establish uncertainty budgets and calculate expanded uncertainties in agreement with the ISO Guide to the expression of uncertainty in measurement.

Under these circumstances, expanded uncertainties of both CRM and laboratory should to a certain extent have to overlap to be acceptable.

To make proper use of this information, ISO Guide 33 should be revised accordingly.

- First, it is questionable if an assessment of precision of a measurement process is still the relevant thing to do, and if it should not be replaced by an assessment of (combined standard) *uncertainty*. In any case, it should be complemented by such an assessment as the real preliminary question to evaluate the acceptability a measurement method is "is the combined standard uncertainty of the measurement process sufficiently fit for purpose?". Obviously, it is not the task of the CRM producer to answer this question, although CRM's certified by laboratory intercomparison may provide useful guidance to the user in clarifying what may be expected or considered as "normal".
- If this criterion is satisfied, the trueness of the measurement process can be assessed within the combined standard uncertainties of the CRM sample and the laboratory measurement. The proposal is therefore to replace σ_D in equation [3] by the combined standard uncertainty (u_C) of the CRM - (u_{CRM}) and the laboratory measurement (u_{meas}) uncertainties, i.e.:

$$u_C = [u^2_{CRM} + u^2_{meas}]^{1/2} \qquad [5]$$

resulting in:

$$- a_2 - 2\,u_C \le \overline{x} - \mu \le a_1 + 2\,u_C \qquad [6]$$

whereby one may ask himself if after replacing σ_D by combined standard uncertainties according to GUM there is still a justification or a need for adjustment values, as proposed in ISO Guide 33.

- whenever possible, minimum representative sample sizes should be determined on experimental grounds and sampling constants made available so that uncertainties can be recalculated when smaller samples are used than those to which the certified value apply (e.g. for calibration of solid sampling spectroscopic methods).

The problem is, however, that - at least for BCR CRM's - changes will not be implemented in practice before applicable guidelines (e.g. BCR, ISO-REMCO) will effectively be adapted, which will require a long and tedious decision making process.

References

1. ISO Guide 33, Use of certified reference materials, ISO, Geneva, 1998.
2. Jorhem, Fresenius J. Anal. Chem., 1998, **360**, 370.
3. ISO Guide 30, Terms and definitions used in connection with reference materials, ISO, Geneva, 1992.
4. Guide to the expression of uncertainty in measurement, ISO, Geneva, ISBN 92-67-10188-9, 1995.
5. The role of reference materials in achieving quality in analytical chemistry, ISO, Geneva, ISBN 92-67-10255-9, 1997.
6. Catalogue of BCR Reference Materials, European Commission, IRMM, Geel, Belgium, 1998.
7. Lamberty and H. Muntau, The certification of the mass fraction of As, Cd, Cr, Cu, Hg, Mn, Pb, Se and Zn in mussel tissue (CRM 278R), EUR-report (submitted for publication).

8. Guidelines for the production and certification of BCR reference materials, doc. BCR/01/97, European Commission, SMT - programme, Brussels, Belgium, 1997.
9. Quantifying uncertainty in analytical measurement, 1st Edition, Eurachem, ISBN 0-948926-08-2, 1995.
10. Lamberty, Ph. Quevauviller and M. Morabito, The recertification of the contents (mass fractions) tributyltin and dibutyltin (CRM 462), EUR-report (submitted for publication)
11. Pauwels, A. Lamberty and H. Schimmel, Accred. Qual. Assur., **3**, 51, 1998.
12. Pauwels, A. Lamberty, P. De Bièvre, K.-H. Grobecker and C. Bauspiess, Fresenius J. Anal. Chem., **349**, 409, 1994.
13. Pauwels, C. Hofmann and K.-H. Grobecker, Fresenius J. Anal. Chem., **345**, 475, 1993.
14. Pauwels and C. Vandecasteele, Fresenius J. Anal. Chem., **345**, 121, 1993.
15. ISO Guide 35, Certification of reference materials - General and statistical principles, ISO, Geneva, 1989.
16. Lamberty, P. De Bièvre and A. Goetz, Fresenius J. Anal. Chem., **345**, 310, 1993.
17. Pauwels, F. Lievens and C. Ingelbrecht, Certification of a uranium-238 dioxide reference material for neutron dosimetry (EC nuclear reference material 501a, EUR-12438 EN, 1989.
18. Lamberty, G.N. Kramer, J. Pauwels, I. Krause and H. Glaeser, The certification of two reference materials to be used for the detection of cows' milk casein in cheeses from ewes' milk, goats' milk and mixtures of ewes' and goats' milk, EUR-17254 EN, 1996.
19. Pauwels, A. Lamberty and H. Schimmel, Accred. Qual. Assur, **3**, 180, 1998.

Proper Use of Reference Materials for Elemental Speciation Studies

K. Okamoto

NATIONAL INSTITUTE OF MATERIALS AND CHEMICAL RESEARCH, 1-1, HIGASHI, TSUKUBA, IBARAKI, 305-8565, JAPAN

J. Yoshinaga

NATIONAL INSTITUTE FOR ENVIRONMENTAL STUDIES, 16-2, ONOGAWA, TSUKUBA, IBARAKI, 305-0053, JAPAN

1 INTRODUCTION

Use of certified reference materials (CRMs) is essential to maintain accuracy and precision of analytical results, particularly in the analysis of real world specimens such as biological and environmental samples. To meet these ever-increasing demands for matrix reference materials, the National Institute for Environmental Studies (NIES) has undertaken over the last 20 years the preparation of a variety of biological and environmental reference materials (RMs). The NIES CRMs for trace element analysis include Pepperbush,[1] Pond Sediment,[2] Chlorella, Human Hair,[3] Mussel,[4] Tea Leaves,[5] Vehicle Exhaust Particulates,[6] Sargasso Seaweed[7] and Rice Flour Unpolished.[8]

Since essentiality and toxicity of an element are closely related to its chemical forms, elemental speciation studies have become increasingly important in human nutrition, environmental sciences and so on. NIES recently launched new projects for the preparation of biological and environmental CRMs for elemental speciation studies. Preparation of biological and environmental matrix reference materials for use in metal speciation studies is much more difficult because, in general, chemical forms of metals may not be stable during sample preparation and in the following long-term storage. Depending on the type of samples and on the chemical forms of the elements to be determined, studies on sample preparation such as pulverization, homogenization, drying, packaging, preservation conditions, *etc.,* should be conducted in order to obtain homogeneous and stable natural-matrix reference materials. Elemental speciation studies invariably require the development of so-called "hyphenated" methods, which combine a separation method and a detection technique in sequence, to determine the individual metal-containing compounds present usually at ppm (mg kg^{-1}) level or less in complex natural matrices.

Fortunately, recent remarkable developments of both separation and detection techniques combined such as gas chromatography-mass spectrometry (GC-MS) and high performance liquid chromatography-inductively coupled plasma mass spectrometry (HPLC-ICP-MS) have made it possible to determine the chemical forms of elements in complicated matrices on a routine basis if standard substances are available. NIES recently issued NIES No. 11 Fish Tissue[9] and NIES No. 12 Marine Sediment[10,11] CRMs for use in the analysis of organotin compounds. NIES No. 13 Human Hair[12] RM was prepared as a renewal of the previous NIES No. 5 Human Hair CRM and is certified for methylmercury (MeHg) and several trace elements. NIES No. 14 Hijiki Seaweed[13] and NIES No. 15

Scallop Tissue RMs have been prepared for inorganic and organic As speciation studies, respectively.

In this paper, as examples of natural matrix reference materials for use in elemental speciation, the preparation and certification of NIES Marine Sediment RM for tin speciation and NIES Human Hair RM for MeHg and trace elements are described, together with additional information for proper use of these CRMs.

2 NIES MARINE SEDIMENT REFERENCE MATERIAL FOR SPECIATION OF ORGANOTIN COMPOUNDS

Marine pollution with organotin compounds, an antifouling agent used on ship hull, has been a matter of great concern in the environmental and fishery sciences. In addition to its high toxicity to a variety of marine organisms, an endocrine disrupting activity of tributyltin (TBT) and triphenyltin (TPT) to a certain species of marine gastropod at a considerably low level (1 ng L^{-1} in seawater) has recently been highlighted. Although the marine environmental level of TBT in Japan has been reported to be decreasing after the regulation of the use of certain TBT compounds started in 1990, the decreasing rate seems to be slowing in the past few years. The environmental and biological monitoring of organotin compounds is, therefore, still of great importance for conserving the marine environment. Analytical quality assurance of organotin determinations is essential to obtain accurate results on relatively low concentrations of organotin compounds in the environmental and biological samples.

In order to meet the demands for new types of RMs, NIES issued Fish Tissue[9] CRM (NIES CRM No.11) with certified values for the TBT and total tin contents together with a reference value for TPT. Fish fillets of sea bass were minced, freeze-dried, ball-milled, homogenized and finally vacuum-packaged, with an oxygen absorbent, into polyethylene laminate bags (600 samples, 20 g each). The TBT and TPT contents in the material were determined by GC-flame photometric detection (FPD), GC-electron capture detection (ECD) and GC-MS. The certified value for TBT was determined to be 1.3 ± 0.1 μg g^{-1} as its chloride and the reference value of 6.3 μg g^{-1} was given for TPT. The total tin content of 2.3 μg g^{-1} was certified based on the results obtained by isotope dilution (ID)-ICP-MS, hydride generation atomic absorption spectrometry (HGAAS). Samples of NIES Fish Tissue CRM has been distributed worldwide on request from many scientists working on the determination of organotin compounds.

Sediment is an important component in the aquatic dynamics of organotin compounds, because the half-life of TBT in sediments is longer compared to that in sea water. Sediments can be considered as a sink of TBT as well as a releasing source after its deposition into the marine environment. The Harbor Sediment[14], PACS-2, issued by the National Research Council of Canada (NRCC) is the only currently available sediment CRM for tin speciation studies, though several sediment CRMs have been issued by NIST, NRCC and BCR. The certified value for TBT in the harbor sediment PACS-2 (TBT 0.98 ± 0.13 mg Sn kg^{-1}) is much higher than those generally found along the Japanese coast. The PACS-2 can be considered suitable for quality assurance of organotin analysis of heavily contaminated sediments. Availability of a marine sediment RM certified for lower levels of organotin compounds will be of value for environment monitoring in background or less contaminated areas. Therefore, NIES has recently undertaken the preparation and certification of a marine sediment RM for tin speciation studies.

2.1 Preparation

The starting material of NIES marine sediment RM was sampled in 1989 at the central area of Tokyo Bay. Approximately 200 kg (wet weight) of the surface sediment was collected with an Eckman-Burge sampler and transported to NIES. The seawater was removed by filtration on a Büchner funnel. The sediment was air-dried for about 2 weeks. After removing visible external materials (*e.g.*, shell), the sediment was ground in a 7L high-purity alumina ball-mill to pass a 100 μm nylon sieve. The <100 μm fraction was homogenized in one lot using a V-blender (200 L) for 3 h. A 30 g aliquot of the homogenized material was doubly vacuum-packaged into a polyethylene laminate bag with oxygen absorbent to prepare 700 packages. Sterilization by Co-60 irradiation was not applied to prevent possible decomposition of the organotin compounds. The samples in a vacuum-packaged form have been stored at −20 °C in a freezer to prevent bacterial activities.

2.2 Homogeneity Assessment

Homogeneity of the prepared material was examined by analyzing major and trace element concentrations in the RM. Three samples (300mg each) were taken from each of 3 randomly selected packages and decomposed with a mixture of $HNO_3/HClO_4/HF$. The major and trace element concentrations (Na, Mg, Al, K, Ca, V, Cr, Fe, Mn, Cu, Zn, and Sr) were determined by inductively coupled plasma atomic emission spectrometry (ICP-AES). The within- and between-package variations were estimated by one-way analysis of variance (ANOVA). The within-package variations for the elements determined were close to the measurement imprecision of each element by ICP-AES, ranging from 0.5 to 1.7% as the relative standard deviation (RSD). The ANOVA showed no statistically significant between-package variation in the elemental concentrations. These results indicate that the prepared material has a homogeneous elemental composition.

Particle size distribution of the prepared material was analyzed with a LA-910 Laser Scattering Particle Size Distribution Analyzer (Horiba Co. Ltd., Kyoto, Japan) after ultrasonic dispersion of the sample into water for 1 min. The particle size distribution showed a normal distribution with a median particle size of 10 μm.[11] This result on the physical property also support excellent homogeneity of the material.

2.3 Collaborative Analysis and Certification

Certification of the prepared NIES marine sediment RM was carried out based on collaborative analysis. All of the results from collaborative laboratories were first checked for completeness of the reporting format. When a part of the information was missing, the collaborator was asked to provide it. Then, the mean and standard deviation were calculated for each set of analytical results. The standard deviation was examined taking into consideration the expected standard reproducibility of each analytical method employed. A Grubbs' test was first applied to reject an outlier mean value, if present, so as to obtain an acceptable mean value. The certified value was determined as the mean of the all acceptable mean values, while the uncertainty range is expressed, in this case, as two times the standard deviation of the mean.

2.3.1 Moisture Content. In order to prevent possible decomposition of organotin compounds, air-drying was applied in the preparation process of this RM. Therefore, this material contains a significant amount of moisture. Because the certified and reference values are expressed on a dry weight basis, a moisture content of this CRM has to be measured and then corrected to relate the analytical values to the certified and reference values. An average moisture content of 6% was obtained under the condition described in Table 2 from repetitive measurements when the new packages were opened at NIES. However, once a package was opened, the moisture content may vary with time depending on the environment. For this type of RMs, it is required to measure the moisture content on a separate sample prior to measuring any analytes of the CRMs if the package had been opened.

2.3.2 Certification of Tributyltin. Nine laboratories participated in a collaborative analysis for certification. Two packages were sent to some of the laboratories to assess the between-package variability by analyzing more than 2 sub-samples from each of the 2 packages. The collaborating laboratories were asked to analyze TBT and other organotin compounds using their routine methods. The analytical methods used in the collaborative analysis included GC-ECD, GC-FPD, GC-MS and HPLC-ICP-MS, preceded by acidic solvent extraction, derivatization, and clean-up.

Figure 1 shows a GC - atomic emission detection (AED) chromatogram of the ethyl-derivatives of dibutyltin (DBT), TBT and TPT extracted from NIES Marine Sediment RM. Tripentyltin (TPeT) and tetrapentyltin (TePeT) were added as a surrogate compound and an internal standard, respectively. There are some other unidentified organotin compounds on the chromatogram.

The 9 collaborating laboratories provided 11 sets of analytical values, each set comprising 2-6 individual values. Table 1 gives a summary of the analytical procedures and the mean of the reported values provided by the collaborating laboratories. A Grubbs' test rejected one of the 11 mean values. The certified value for the TBT content of NIES CRM No. 12 Marine Sediment was determined to be 0.19 ± 0.03 mg kg^{-1} dry weight, based on the mean and 2 times the standard deviation of the ten values. The certified value is listed in Table 2. Note that the certified value is expressed on a TBT ion $[(CH_3CH_2CH_2CH_2)_3Sn^+]$ basis.

In order to examine stability of TBT and TPT in this RM, two independent determinations of TBT and TPT were performed by GC-MS at NIES with a 3-year interval using newly opened packages each time. No detectable difference was found between the two determinations. It is thus assumed that TBT and TPT in this CRM is stable under the specified storage condition. However, the stability of TBT and other constituents has not been monitored once the package was opened or in case the package was stored under different conditions.

2.3.3 Reference Value for Triphenyltin. Seven laboratories provided analytical values for TPT in Marine Sediment RM. However, the reported mean values ranging from 0.006 to 0.010 mg TPT kg^{-1} dry weight made it difficult to determine a certified

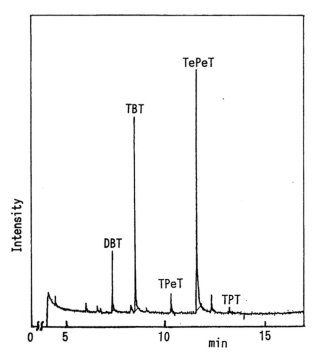

Figure 1 *A gas chromatograph-atomic emission detection chromatogram of dibutyltin (DBT), tributyltin (TBT) and triphenyltin (TPT) in the extract of NIES Marine Sediment. Tripentyltin (TPeT) and Tetrapentyltin (TePeT) were added as a surrogate compound and an internal standard, respectively.*

Table 1 *Summary of Collaborative Analysis of Tributyltin in NIES Marine Sediment RM*

Lab No	Mean of Reported values [a]	Extraction	Derivatization	Analytical Method [b]
1	0.20	HCl/methanol-ethyl-Acetate/hexane-ether	None	GC-ECD
2	0.12 [c]	HCl/ethylacetate	Hydride	GC-FPD
3	0.20	HCl/dichloromethane	Ethyl	GC-FPD
4	0.17	NaBEt$_4$-acetate buffer	Ethyl	GC-MS
5	0.190	HCl/methanol-ethylacetate	Ethyl	GC-MS
	0.200	HCl/methanol-ethylacetate	Ethyl	GC-MS
6	0.187	HCl/acetone/tropolon-benzene	Propyl	GC-FPD
7	0.215	HCl/methanol	None	HPLC-ICP-MS
8	0.170	HCl/methanol- ethylacetate	Propyl	GC-MS
	0.184	HCl/methanol-ethylacetate	Propyl	GC-FPD
9	0.170	HCl/methanol	Propyl	GC-FPD

a, expressed as mg TBT kg^{-1} on a dry weight basis, b, for abbreviations, see text, c, rejected by Grubbs' test ($p<0.05$).

value. Since TPT, as well as its endocrine disrupting activity, is an essential component to be monitored in the marine environment monitoring program in our country, it is of great importance to provide an information value of the TPT concentration. In this case, the mean of the reported values, 0.008 mg kg^{-1} expressed as a TPT ion, is given as the reference value (Table 2). Further accumulation of TPT analytical results is required for the certification.

2.3.4 Certification of Total Tin. Accurate determination of total tin is important in geochemistry, environmental sciences, human nutrition, *etc,* because of difficulties in determining trace levels of tin. Six laboratories participated in a collaborative analysis for the certification of total tin content of this RM. The analytical methods used included instrumental neutron activation analysis (INAA), HGAAS, electrothermal AAS (ETAAS), ICP-MS and ID-ICP-MS, preceded by HNO_3 / $HClO_4$ / HF digestion with or without solvent extraction. At NIES, alkaline fusion was also employed for decomposition of the sediment for ETAAS, flow-injection ICP-MS and ID-ICP-MS determinations. Ten sets of analytical values were obtained from the 6 collaborative laboratories.

The certified value for the total tin content was determined based on 10 sets of analytical values reported from collaborating laboratories. Five laboratories used acid digestion for the decomposition of this material, except one laboratory which used INAA. Since it is known that some forms of tin in geological materials are acid-insoluble, $LiBO_2$ fusion was employed at NIES as a pretreatment for the ICP-MS, ETAAS and ID-ICP-MS determinations. No difference was detected between the results obtained by acid digestion and those by alkali fusion, indicating that tin in this RM is present in an acid-soluble form. The results obtained by both decomposition procedures were included in the certification process. A Grubbs' test did not reject any of the results for the total tin. The certified value for the total tin content of NIES CRM No. 12 Marine Sediment was determined as 10.7 ± 0.4 mg kg^{-1} on a dry weight basis, as shown in Table 2.

2.4 Evaluation of Organotin Levels in NIES CRM No. 12 Marine Sediment

To date, five marine sediment RMs were prepared for organotin analysis by NRCC (PACS-1, PACS-2), BCR (CRM 462, CRM 424) and NIES (No.12). However, three of the CRMs (PACS-1, CRM 424, CRM 462) have already been out of stock and are no longer available. The starting material of PACS-2 was a marine harbor sediment, where TBT contamination is, generally speaking, more severe compared to other marine environment; this may be the reason for the higher TBT content of PACS-2 CRM than that of NIES Marine Sediment.

Since 1986 the Environment Agency of Japan has been conducting a biological and environmental monitoring program on TBT and TPT pollution in Japan. According to the results compiled in 1997, the TBT level in marine sediments collected from 36 stations along the Japanese coast ranged from <0.0005 to 0.83 mg TBT kg^{-1} on a fresh weight basis with a geometric mean of 0.010 mg kg^{-1}, while the TPT levels between <0.001 and 0.20 mg kg^{-1} with a mean value of 0.0013 mg kg^{-1} were reported. The TBT and TPT levels in the sediment of a monitoring station in Tokyo Bay, which is very close to the sampling site of the NIES Marine Sediment RM, were reported in 1996 to be 0.11

Table 2 *Certified and Reference Values for NIES No.12 Marine Sediment CRM*

Certified Value [a]		Unit [b]
Tributyltin	0.19 ± 0.03	mg TBT kg^{-1}
Total tin	10.7 ± 1.4	mg Sn kg^{-1}
Reference value		
Triphenyltin	0.008	mg TPT kg^{-1}

a, Mean and 2 times the standard deviation of the all acceptable means provided by collaborating laboratories. b, on a dry weight basis. Weighed sample should be dried at 110 °C for 4 h and cool to room temperature in a silica gel desiccator for 30 min. Analytical values should be corrected for the moisture content to relate to the certified and reference values.

and 0.004 mg kg^{-1} as their chlorides, respectively. Therefore, the TBT and TPT concentrations in NIES CRM No.12 Marine Sediment are higher than the current average levels of organotins in marine sediments in Japan, reflecting the fact that the original material for this CRM had been sampled from Tokyo Bay, one of the busiest water transportation areas in Japan, before the regulation of the use of certain TBT compounds started. However, the organotin levels in this CRM still fall within the range of the currently reported concentrations of organotins in Japan. NIES CRM No. 12 Marine Sediment will be of great use for quality assurance of organotin analysis of moderately contaminated sediments.

3 NIES HUMAN HAIR REFERENCE MATERIAL FOR ETYLMERCURY AND TRACE ELEMENTS

Hair is a suitable medium for monitoring human exposure to Hg. Mercury levels in hair reflect organ Hg levels and dietary intake. In addition to analysis of total Hg, a significance of Hg speciation in environment and biological systems has been stressed for a long time, particularly after experiencing population-level poisonings by organic Hg species in several parts of the world. Among the organomercury species, particular concern has been focused on methylmercury (MeHg) because of its ubiquitous presence in nature and its extreme toxicity to humans.

The demand for a human hair CRM for Hg speciation has thus been increasingly becoming important. Although human hair CRMs are currently available from the Community Bureau of Reference (BCR, EU) and the Shanghai Institute of Nuclear Research (China), MeHg concentrations are not certified in these CRMs (an information value is given for the BCR CRM). The International Atomic Energy Agency (IAEA) recently prepared a human hair powder for use in an interlaboratory comparison for MeHg.[15] In 1985, NIES issued the No. 5 Human Hair CRM, prepared from male scalp hair.[3] This RM was certified for Ca, Cd, Cr, Cu, Fe, Hg, K, Mg, Mn, Na, Ni, Sr and Zn, together with reference values for Al, Ba, Br, Cl, Co, P, Pb, Rb, Sb, Sc, Se and Ti, but unfortunately samples of NIES No.5 are now out of stock. In consideration of the great demand for a human hair CRM to assess the toxicological and nutritional status of individuals, NIES recently prepared a new human hair CRM for the determination of MeHg and several trace elements. Special attention was paid to reduce contamination from a grinding vessel to an insignificant level so that the prepared material can be representative of human hair having normal concentration ranges of elements.

3.1 Preparation

Male scalp hair used for the preparation of this new hair RM was the same as that used for the previous NIES human hair CRM (No. 5). The hair was collected in three barbershops in Tsukuba and Tokyo in 1980. Ten kg of the stock human hair was used. Approximately 10 kg of the hair from the stock was divided into several portions, and each portion was ultrasonically washed with non-ionic detergent (0.4 % poly-oxyethylenelaurylether) in a nylon bag. After vigorous rinsing with a large volume of distilled water, the hair was dried on a polypropyrene tray at 50 °C overnight.

A ceramic disc mill made of silicon nitrite (Si_3N_4) was developed in order to minimize contamination. The hair was pulverized with the disc mill, which consists of an inner ceramic (silicon nitrite) ring, a Teflon intermediate ring and an outer ceramic vessel. A 5 – 10 g portion of the hair was put into the spaces between the outer and intermediate rings, and between the intermediate and the inner rings. Liquid nitrogen was then poured into the mill and, after the violent boiling ceased, a ceramic lid was placed onto the disc. The disc mill was placed in a Shatterbox 8510 (Spex Industries, Inc., NJ, USA) to perform cryogenic-grinding for 3 min. The pulverizing efficiency was satisfactory; approximately 10 g hair was reduced to fine powder, about 30% of which passed a 100 μm mesh screen. The hair powder thus obtained was stored in polyethylene bags. The most serious technical problem of this pulverization technique is that only 10 g of hair could be treated in one operation. Contamination of elements from the disc mill will be discussed later.

The hair powder was passed through a nylon screen to obtain approximately 3 kg of hair powder of <100 μm mesh size. The sieved powder was then put into a borosilicate barrel (10 L) in one lot and blended for 8 h using a Rocking Mixer (Asahi Rika Garasu Kogyo, Nagoya, Japan). The homogenized hair powder was packaged, 3 g each, into 1,000 precleaned borosilicate glass bottles. Sterilization by gamma-ray irradiation was not applied to avoid possible decomposition of MeHg. The samples of the human hair powder have been stored at –20 °C in the dark to reduce microbial activity and photodegradation.

3.2 Homogeneity Assessment

From the lot of 1,000 bottles, 5 bottles were randomly selected for a homogeneity test. Five samples (20 to 30 mg each) taken from each bottle were directly analyzed for total mercury content using pyrolysis-gold amalgamation-atomic absorption spectrometry (PAAS) with a MA-IS/MD-1 mercury analyzer (Nihon Instruments Co. Ltd., Osaka, Japan). Another 5 samples (approximately 120 mg each) taken from each bottle were digested with nitric acid by the Teflon double digestion vessel method.[16] The element contents were determined by ICP-AES with an ICP-750 spectrometer (Nihon Jarrell-Ash, Kyoto, Japan) and by air-acetylene flame atomic absorption spectrometry (FAAS) with a SAS-760 spectrometer (Seiko Instruments Co. Ltd., Tokyo, Japan).

The analytical results on the total Hg showed that the between- and within-bottle variations expressed as the relative standard deviation (RSD, %) were 0.4 and 0.9 %, respectively. The analysis of variance revealed no statistically significant variation between the bottles ($F=0.213$, $p=0.928$). These results indicate that the prepared RM is practically homogeneous for the total Hg content for 20 to 30 mg samples.

Among the elements tested for homogeneity by ICP-AES (Ca, Cu, Fe, Mg, Mn, P, Sr and Zn) and FAAS (Na and K) measurements after dissolution with nitric acid, the between-bottle variation for Mg was found statistically significant. The within-bottle variations for Fe and Mn contents, whose between-bottle variations showed no statistical significance, were also found to be larger than the imprecision of the ICP-AES measurements. Contamination from the disc mill may be a possible explanation of the variation. In consideration of these results Fe, Mg and Mn were excluded from the elements to be certified. With respect to other elements examined, the within-bottle variations were close to the imprecision of the ICP-AES and AAS measurements. In addition, no statistically significant variations between the bottles were found for Ca, Cu, K, Na, P, Sr and Zn, indicating good homogeneity of the prepared RM for these elements for 120 mg samples.

The use of a mixture of nitric acid and hydrofluoric acid for the digestion of the prepared human hair RM improved the within-bottle variations for elements such as Al and Fe. This fact suggests that the prepared RM contains a very small quantity of silicates in which these elements are contained, though no visible insoluble matter was observed in the nitric acid-digested solution. Therefore, the addition of hydrofluoric acid is required for the complete digestion of the prepared RM.

3.3 Certification

Certified values for MeHg, total Hg, Cd, Cu, Pb, Sb, Se, and Zn were determined based on analytical results obtained by at least three independent analytical method. The principle of the certification process was adopted from Okamoto.[1] Table 3 indicates the analytical methods used for each element certified.

3.3.1 Methylmercury. Five laboratories provided analytical results on MeHg obtained by GC-ECD following HCl-toluene/benzene-extraction and clean-up procedure. The mean value $\pm$ 95% confidence interval was 3.85 ± 0.28 $\mu g\ g^{-1}$ as Hg on a dry weight basis. Data obtained by aqueous phase ethylation-GC-cold vapor atomic fluorescence spectrometry (CVAFS) (3.61 $\mu g\ g^{-1}$) after alkali digestion and by extraction-PAAS (3.62 $\mu g\ g^{-1}$) agreed well with the GC-ECD values. At NIES, MeHg was determined by a reversed-phase HPLC coupled with an ICP-MS detection system. Methylmercury was extracted by heating (95 °C) the sample in 2 M HCl for 5 min followed by toluene extraction and back-extraction into HPLC mobile phase (0.02 M cysteine in 0.1 M acetic acid, pH 2.2). The mean value of 3.82 $\mu g\ g^{-1}$ with the standard deviation of 0.12 $\mu g\ g^{-1}$ was obtained for 5 samples. The certified value for MeHg was determined to be 3.8 ± 0.4 $\mu g\ g^{-1}$ as Hg. No other organomercury species were detected by HPLC-ICP-MS in the toluene extract of the prepared human hair RM. No collaborationg laboratories detected the presence of other organomercury species in the material. Thus MeHg can be considered to be the only form of organomercury (toluene or benzene extractable) species present in NIES Human Hair CRM, justifying the use of extraction-PAAS data in the certification of MeHg.

3.3.2 Total Mercury. Four laboratories supplied analytical values on total Hg by INAA. The mean value $\pm$ the 95% confidence interval for the mean was calculated to be 4.47 ± 0.18 $\mu g\ g^{-1}$ (dry weight basis). The PAAS data were provided by three laboratories with the mean $\pm$ the 95% confidence interval of 4.44 ± 0.07 $\mu g\ g^{-1}$. Data derived from cold vapor AAS (4.30, 4.36 $\mu g\ g^{-1}$) and from CVAFS (4.44 $\mu g\ g^{-1}$) measurements were all in good agreement with those obtained by INAA and PAAS. At NIES, ID-ICP-MS was applied to determine total Hg in the human hair. Briefly, an aliquot of ^{202}Hg was spiked to a sample and the mixture was digested with nitric acid by the Teflon double vessel digestion method.[16] The altered ^{200}Hg/^{202}Hg ratio was then measured with an ICP mass spectrometer to calculate the Hg content. The analytical value by ID-ICP-MS was 4.31 ± 0.08 µg g^{-1} Hg (n=5), showing good agreement with those determined by other analytical techniques. Based on these values mentioned above, the certified value for total Hg was determined to be 4.42 ± 0.20 µg g^{-1}.

3.3.3 Other Trace Elements. Extensive analyses of heavy metals such as Cd, Cu, Pb, Sb, Se and Zn in human tissues have been carried out for nutritional and toxicological studies. Most biological and clinical CRMs lack certified values for so-called "difficult elements" which may be essential for animal growth but are present at very low levels (ng g^{-1} or less) and are thus difficult to accurately determine. Therefore, availability of certified values for these elements in human tissue RMs will be of great value to study essentiality and toxicity of the elements.

Table 3 lists the certified values for Cd, Cu, Pb, Sb, Se, and Zn in the human hair CRM, together with the analytical methods used for the determination of the elements. At least 4 independent analytical techniques were employed for each element certified in the collaborative analysis by 5-11 participating laboratories. Analytical results by ICP-MS and microwave induced nitrogen plasma mass spectrometry (MIP-MS) coupled with isotope dilution technique were provided for Cd, Pb, Sb, Se and Zn, in addition to the total Hg, from 3 different laboratories. Certification of these elements was performed according to the procedure described above. The certified values for NIES No.13 Human Hair are shown in Table 3.

3.3.4 Reference Values. Reference values are given for Ag, Al, As, Ba, Ca, Co, Fe, Mg, Mn, Na, S, and V, for which analytical results were obtained from only one or two independent analytical methods (but the agreement between different laboratories was satisfactory, *e.g.*, Co) or for which the between-bottle variations were statistically significant (*e.g.*, Fe and Mg). The reference values include some difficult elements of toxicologically interests. Availability of further analytical data for these elements will enable some reference values to be certified.

3.4 Elemental Composition of NIES No.13 Human Hair CRM

The elemental concentration of the prepared NIES No.13 Human Hair is similar to that of NIES CRM No. 5 Human Hair, because these two CRMs were prepared from the lots of the same stock of Japanese male scalp hair. NIES No. 13 Human Hair has lower concentrations of Al, Fe and Mg compared with No. 5 as a result of the use of the ceramic disc mill for pulverization. However, the levels of these elements in NIES No.13 are still higher than those reported as the normal concentration ranges for the

Table 3 *Certified and Reference Values for NIES No.13 Human Hair CRM*

Certified value [a]		Reference value [a]			
Total Hg	4.42±0.20	Al	120	Mn	3.9
Methyl Hg [b]	3.8±0.4	Ag	0.10	Na	61
Cd	0.23±0.03	As	0.10	S(%)	5.0
Cu	15.3±1.3	Ba	2.0	V	0.27
Pb	4.6±0.4	Ca	820		
Sb	0.042±0.008	Co	0.07		
Se	1.79±0.17	Fe	140		
Zn	172±11	Mg	160		

a, µg g^{-1} on a dry weight basis unless otherwise indicated. Samples should be dried
at 85 °C for 4 h followed by cooling in a silica gel desiccator

b, expressed as µg Hg g^{-1} on a dry weight basis.

Japanese male, suggesting that contamination from the grinding vessel was still significant
for the elements. According to the manufacturer, some trace elements including Fe and Mg
are added as sintering additives into the silicon nitrite. The Se concentration in NIES
No.13 Human Hair is slightly higher than a normal concentration range, probably due to
pretreatment of the hair sample, *e.g.,* use of a certain medication containing Se.

Except for the above-mentioned elements, the elemental concentration of NIES No.13
Human Hair CRM can be considered similar to the normal concentration range of the
elements for the Japanese male.

References

1.   K. Okamoto, Res. Rep. Natl. Inst. Environ. Stud., 1980, **18**, 20
2.   K. Okamoto, Res. Rep. Natl. Inst. Environ. Stud,. 1982, **38**, 1
3.   K. Okamoto, M. Morita, H. Quan, T. Uehiro and K. Fuwa, *Clin. Chem.*, 1985, **31**, 1592
4.   K. Okamoto and K. Fuwa, *Analyst*, 1985, **110**, 785
5.   K. Okamoto and K. Fuwa, *Fresenius Z. Anal. Chem.*, 1987, **326**, 622
6.   K. Okamoto, *Anal. Sci.*, 1987, **3**, 191
7.   K. Okamoto, *Marine Environ. Res.*, 1988, **26**, 199
8.   K. Okamoto, *Sci. Total Environ.*, 1991, **107**, 29
9.   K. Okamoto, 'Biological Trace Element Research', Am. Chem. Soc., Washington, DC, 1991, ACS Symposium Series No.445, Chapter 20, p.257.
10. J. Yoshinaga, A. Tanaka, T. Takamatsu, M. Morita and K. Okamoto, *Anal. Sci.,* 1996, **12**, 993
11. J. Yoshinaga, H. Kon, T. Horiguchi, M. Morita and K. Okamoto, *Anal. Sci.*, in press
12   J. Yoshinaga, M. Morita and K. Okamoto, *Fresenius J. Anal. Chem.*, 1997, **357**, 279
13. K. Okamoto, J. Yoshinaga and M. Morita, *Mikrochim. Acta*, 1996, **123**, 15
14. Certificate of Analysis of PACS-2, NRCC, Canada, 1998
15. S. F. Stone, F. W. Backhaus, A. R. Byrne, S. Gangadharan, M. Horvat, K. Kratzer, R. M. Parr, J. D. Schladot and R. Zeisler, *Fresenius J. Anal. Chem.*, 1995, **352**, 184
16. K. Okamoto and K. Fuwa, *Anal. Chem.*, 1984, **56**, 1758

NIST Standard Reference Materials for Measurement Assurance – Practices, Issues and Perspectives

Thomas E. Gills

STANDARD REFERENCE MATERIALS PROGRAM, GAITHERSBURG, MD 20899, USA

1 INTRODUCTION

Standard Reference Materials (SRMs) are certified reference materials (CRMs) issued under the NIST trademark that are well-characterized using state-of-the-art measurement methods and/or technologies for chemical composition and/or physical properties. Traditionally, SRMs have been the primary tools that NIST (formerly NBS) provides to the user community for achieving measurement quality assurance and traceability to national standards. Each SRM is the result of collaboration between NIST and representatives of science and industry. SRMs are crucial reference points in establishing a comprehensive measurement system for the U.S.

SRMs are designed to enable analysts to check or calibrate their entire measurement system within their laboratory by providing them well-characterized materials and the analytical results of the characterization procedures detailed in the SRM Certificate. The need to ship equipment to a central calibration point or have inspection teams come to the laboratory facility is eliminated by the availability of SRMs. However, with the fast pace of technological change, coupled with increased demands on quality, traceability, and needs for SRMs of varied types, the NIST faces a very challenging proposition in providing the user with the breath and number of needed and often required SRMs. Consequently, the user or potential user of SRMs is often faced with the problem of unavailability of SRMs or reference materials. Alternative sources of RMs or secondary standards whose values and uncertainties on different but similar materials may not reflect compatibility in measurement further complicate the problem. The matrix and form of the SRM needed will not always be available due to the lack of available resources to develop, produce, and characterize in a timely manner. To SRM/RM users, these important issues are foundational to the decisions that they must make in both the selection and proper use of SRMs. This presentation will cover the NIST certification process; process of selecting appropriate SRMs, CRMs, or RMs; SRM traceability hierarchy; SRM maintenance and storage problems; and availability information.

2 CERTIFICATION PROCESS

There are two aspects of SRM certification, one legal, the other scientific. Legally, the certification process indicates that a NIST SRM carries the full weight and legal authority of both the U.S. Department of Commerce and NIST, in that, these are official materials authorized by appropriate federal laws and regulations. NIST SRMs are incorporated into many regulatory requirements as measuring tools for assuring quality and for achieving traceability to the U.S. National Measurement System developed and maintained by NIST. The scientific aspects of certification are of primary importance to the analyst. The issuance of a SRM/RM is guided by an experimental plan that incorporates an assessment of the measurement need, material type and physical form, the required measurement results, and statistical criteria for sampling and analysis. The experimental plan defines the following key components of the certification process:

- Required properties of the base material and SRM/RM unit
- Statistical plan for subsampling and measurement
- Required scope of measurement data and limits of uncertainty
- Techniques and methods to be used in the measurement process
- Measures of traceability and verification
- Provision for maintaining the SRM / RM (stability tests) and propagation of the SRM / RM

The goal of any SRM certification is to report the "true value" of a given property(ies) under investigation and the level of confidence in the "true value". Therefore, NIST measurement results are accompanied by quantitative statements of uncertainty [1]. To ensure that such statements are consistent with each other and with present international practice, the NIST policy for evaluating and expressing the uncertainty adopts in substance the approach recommended by the International Committee for Weights and Measures (CIPM) [1,2].

How does NIST value assign its SRMs? Historically, NIST has used three basic modes: a) measurement by method(s) of high precision for which sources of bias have been rigorously investigated. The applicability of the method(s) have been demonstrated and documented across a range of diverse matrices; b) measurement by two or more independent and reliable methods whose estimated uncertainties are small, relative to the accuracy required for certification or SRM purpose and use; and, c) measurement via a network of qualified laboratories [3].

In response to increased customer needs for documentation and clearly defined terms associated with assigned values of reference materials and the value-assignment process, NIST has expanded its basic value assignment modes to eight modes for chemical measurements. The eight modes are given below in **Table 1**, along with defined values, i.e. certified values, reference values, and information values. The basic principles remain unchanged , however, NIST's eight defined modes provide a clearer link between the nature of the value-assignment process and the definition of the assigned value (e.g. certified value, reference value, information value). The strength of the measurement link is directly related to the mode(s) used in the value-assignment process.

Within the eight defined modes, there are three basic modes for value-assignment that lead to certified values and five value-assignment modes that provide reference values and/or information values. <u>A NIST Certified Value</u> represents data reported on an SRM Certificate for which NIST has the highest confidence in its accuracy in that all known or suspected sources of bias have been fully investigated or accounted for by NIST. <u>A NIST Reference Value</u> is a best estimate of the true value provided on a NIST Certificate, Certificate of Analysis, and/or Report of Investigation where all known or suspected sources of bias may not have been fully investigated by NIST. <u>A NIST Information Value</u> is considered to be a value that will be of interest and use to the SRM/RM user, but insufficient information is available to assess the uncertainty associated with the value. [4].

Table 1 *Modes Used at NIST for Value-Assignment of Reference Materials for Chemical Measurements*

		C	R	I
1.	Certification at NIST Using Primary Method with Confirmation by Other Methods	☑	☐	☐
2.	Certification at NIST Using Two Independent Critically Evaluated Methods	☑	☑	☐
3.	Certification/Value Assignment Using One Method at NIST and Different Methods by Outside Collaborating Laboratories	☑	☐	☐
4.	Value Assignment Based on Measurement of Two or More Outside Collaborating Laboratories Using Different Methods	☐	☑	☑
5.	Value Assignment Based on a Method Dependent (procedurally-defined) Technique	☐	☑	☑
6.	Value Assignment Based on NIST Measurements by a Single Method (but does not meet criteria for certification)	☐	☑	☑
7.	Value Assignment Based on Outside Collaborating Laboratory Measurements Using a Single Method	☐	☑	☑
8.	Value Assignment Based on Selected Data from Interlaboratory Studies	☐	☑	☑

<u>Key</u>
C = Certified Value
R = Reference Value
I = Information Value

3 APPROPRIATE SRM

Basically, SRMs have three possible uses. They may be used as *control* materials analyzed simultaneously with unknown materials; as *calibration* materials to calibrate instruments or equipment; or as *known* materials used in the development of new methods or techniques. In each instance , the SRM provides the results the analysts should obtain. Therefore the analyst is provided with the means for a critical examination of his measurement process.

When an SRM is used as a *control* material , the employed measurement method may be biased because of the matrix effects. To eliminate or minimize the effects of matrices on the results, the SRM used for the application should meet the following requirements to the extent possible:
1) Reasonable matrix match with the materials customarily analyzed,
2) Reasonable match of analyte concentration (s),
3) Uncertainty of the certified concentrations should be small with respect to the requisite uncertainty for the intended use.

The first major use of *calibration* SRMs was in the preparation of emission spectroscopic curves in the early 1940's. These curves were strictly empirical in nature and had to be established by relating instrumental response to a series of SRMs. The response or shape of the curve was affected by a variety of matrix effects, that included matrix effects and metallurgical properties of the material. The impetus for SRMs to generate such curves came from the steel industry. The need was two-fold, to standardize analytical results from laboratory to laboratory and from run to run, secondly to reduce the time required to analyze samples accurately so that analyses could be made quickly (< 5 min) during production processes. The problems encountered in the steel industry are also being faced today by many segments of technology, i.e., the use of rapid, relative methods, to produce accurate analyses of a large number of samples in a relatively short period of time. Therefore, the use of SRMs for calibration work is of particular use to a large and growing segment of both the scientific and technical communities because of the widespread use of instrumentation.

The use of SRMs for *method development* is to assure that a common material containing known values for the chemical or physical properties of a material will permit researchers in different locations to cooperate in the development of new methods. For example, the development of the reference method for calcium in serum was greatly simplified because a clinical SRM for calcium existed-SRM 915, Calcium Carbonate. Currently, NIST issues a number of primary clinical SRMs for this purpose and in other areas. The constituents certified in these types of materials are measured by several different independent methods or techniques and possible interferences and biases are identified. Such certification makes these SRMs invaluable for the comparison of various measurement methods or techniques.

4 REFERENCE MATERIALS SELECTIONS

Regardless of the end use, several factors must be considered in the selection of an appropriate SRM /CRM or RM. Primary standards, SRMs/CRMs, serve a specific need, but should not be used on a daily routine basis. Rather, they should be used to ascertain the quality of secondary standards that are to be used daily. Therefore, the user must decide whether the intended use requires a primary standard or whether a secondary standard would suffice. Secondary standards are available from a variety of sources. Commercial entities, study groups, individual laboratories, or groups of laboratories may prepare and issue these standards. Ideally, these standards should be related to and measured against the primary standards. Obviously, each time a measurement uncertainty is transferred along a measurement infrastructure, an additional uncertainty should be added. Therefore the more measurement levels between the primary and secondary standard, the greater the uncertainty will be.

The use of primary standards issued by a recognized body and the use of secondary standards prepared by commercial entities strengthens the whole measurement system. Experience at NIST has shown that no one organization can possibly supply all of the reference materials needed by one nation. Therefore a national measurement system should include a recognized standards laboratory that produces primary standards and another level of standards production that prepares secondary standards that are directly related to the primary standards. The availability of lower-cost secondary standards promotes wider use of standards. Such availability reduces the demands on the national standards laboratories or CRM producers to be the major supplier of all standards. This permits the national standards laboratories or CRM producers to devote more of its resources to the development of new and other needed primary standards.

5 TRACEABILITY

Traceability involves both legal and technical aspects. NIST is the U.S. metrology laboratory and is responsible for defining and supporting the U.S. National Measurement System infrastructure. NIST standards (artifacts and data) are made available to support commerce, industry, government, and research in their compliance with traceability requirements (when traceable measurement results are required).   SRMs are key components of a laboratory's tools for obtaining accurate measurement results.

There are two components of traceability. The first component is the traceability of NIST measurements to the SI unit (vertical) and secondly   traceability to linkage to other national measurement institutes or organizations (horizontal). The vertical aspects are defined by specific pathways, e.g., modes of certification and value assignment. The vertical component of traceability is maintained by NIST as the national link or keeper of SI units in the U.S. for dealing with both national and international (i.e., national laboratories from other countries) policy issues related to traceability. The horizontal traceability is usually defined by Memoranda of Understanding with national measurement institutes or organizations of a given country.

The goal of a user is to validate a measurement process or system for purposes of calibration, control, research, and/or compliance. Such validation is achieved when the measurement results and associated uncertainties can be linked to some known and recognized standard or stated reference and their associated uncertainties. When such linkage is demonstrated traceability is in theory achieved. In practice, such traceability can be achieved by demonstrating the linkage to the certified values of NIST SRMs (i.e., national standards). The following three tenets are important to the traceability process:
1) Results obtained are comparable (precise) and compatible (accurate, in agreement) with other results made under similar measurement objectives.
2)   The statistical means of the results obtained by the user for NIST SRMs fall within the uncertainty interval (s) of the certified values, and
3)   The uncertainty interval (s) of the measurement results demonstrate statistical control of the measurement process.

6 MAINTENANCE AND STORAGE OF SRMS

NIST distributes many SRMs that pose storage and shipment problems, but which are adequate for the intended purpose if the SRM is properly handled and stored upon receipt. Some of the SRMs include frozen serum, coals which tend to hydrate, solutions in polyethylene bottles which transpire through the wall of the bottles, x-ray film that is affected by humidity, etc. Careful laboratory storage and sample treatment of these materials is usually required to ensure that the certified value can be obtained. If such storage and treatment requirements are thoroughly researched by the producer, there should be no added problem for the user. Furthermore, in cases where stability due to extreme storage conditions would be suspect, NIST initiates the appropriate stability testing program to address the issue and report any substantial findings or results to the users.

Proper inventory storage conditions and the packaging for shipment do however pose difficult problems for a standards producer that should be solved to make a standard reliable and generally available. The nature of the standard must always be considered at its inception, i.e., its stability under specified storage conditions and over a pre-determined time period. With many new processing techniques and packaging materials available, a variety of ways to solve storage and shipment problems exist. However, periodic testing of inventory stock provides an added measure of assurance to the integrity of the standard and its use in quality assuring the laboratory's measurement process. For, example, The concentration of nitric oxide in nitrogen is routinely monitored by NIST by testing lot control cylinders maintained at NIST over many years under laboratory storage conditions. Nitric oxide is highly reactive and cylinder walls of its containment vessel must be specially treated and conditioned before use. However, with such precautions, NIST nitric oxide SRMs occasionally show losses in concentrations over time. Typical results of nitric oxide in nitrogen for SRM 1684b are presented below in **Table 2.** When test results fall outside of the certified value and the associated uncertainty, certificates are revised and customers are promptly notified.

Table 2 *Analysis of SRM 1684b Lot Standards*

Lot Standard Sample #	Lot Standard Cylinder #	Certified NO Conc. (μmol/mol)	Current (9/98) NO Conc. (μmol/mol)	Percent Diff.
44-QL-01 (1998)	AAL068132	New Issue	97.96 (0.5)	New Issue
44-QL-02 (1998)	AAL068152	New Issue	97.99 (0.5)	New Issue
44-1-PL (1994)	ALM034597	97.65 (0.5)	97.50 (0.5)	-0.15
44-2-PL (1994)	ALM034587	97.77 (0.5)	97.56 (0.5)	-0.21
44-51-OL (1991)	ALM003621	95.49 (1.1)	95.61 (1.1)	0.13
44-55-NL (1991)	ALM003605	94.98 (1.1)	95.04 (1.1)	0.07
44-52-ML (1988)	AAL020691	94.35 (1.1)	94.11 (1.1)	-0.25
44-53-KL (1987)	AAL020683	96.91 (1.1)	96.88 (1.1)	-0.04
44-51-KL (1987)	CC52381	93.48 (1.1)	93.41 (1.1)	-0.08
44-29-IL (1982)	AAL009217	93.22 (0.7)	93.18 (0.5)	-0.04

7 INFORMATION ON NIST SRMS

Information on NIST SRMs can be obtained from the following sources: SRM Certificates; NIST internet web sites; SRM Catalogs and Catalog Supplements (web sites and printed copy); A series of NIST Special Publications (260 Series); brochures on selected categories of SRMs; individual announcements on individual SRMs. A listing of NIST SRMs by technical categories is given below. In addition, papers on SRMs and the

SRM Program itself are presented in technical publications and scientific meetings in the U.S. and other countries by NIST technical staff.

Table 3 *NIST SRMs by Categories ; Total Number of SRMs = 1,299*

Industrial Materials
- Ferrous Metals (174)
- Nonferrous Metals (128)
- Microanalysis (6)
- High Purity Materials (190)
- Ceramics and Glasses (30)
- Cements (13)
- Engine Wear Materials (25)

Environmental
- Inorganics (19)
- Organics (36)
- Primary Gas Mixtures (89)
- Fossil Fuels (49)
- Geological Materials and Ores (56)

Physical Properties
- Ion Activity (29)
- Polymeric Properties (25)
- Thermodynamic Properties (64)
- Optical Properties (23)
- Electrical Properties (14)
- Metrology (38)
- Ceramics and Glasses (30)
- X-Ray Spectrometry (12)

Health/Clinical/Foods
- Health and Industrial Hygiene (68)
- Food and Agriculture (41)

Engineering
- Sizing (21)
- Surface Finish (17)
- Nondestructive Evaluation (4)
- Fire Research (6)
- Misc. Performing Eng. Matls. (21)

Radioactivity
- Radiation Dosimetry (2)
- Radioactive Solutions (36)
- Radiopharmaceuticals (13)
- Alpha Particle Point Source (2)
- Carbon-14 Dating (1)
- Accelerator Mass Spectrometry (1)
- Gamma Ray Point Sources (6)
- Radon Emanation (1)
- Natural Matrix Materials (9)

8 CONCLUSION

NIST has met the reference materials needs of the U.S. industry and commerce for nearly 100 years. While the NIST Standard Reference Materials Program has been focused primarily on U.S. needs and requirements, it is clear that these materials address international measurement needs as well. As demonstration of quality and traceability for chemical and physical measurements have become increasingly global issues, the need for internationally recognized and accepted CRMs have increased correspondingly. Thereby, it is important that producers of CRMs continue to promote the proper and most effective uses of CRMs which inevitably will assure compatibility in measurements between nations, industries, and buyer and seller, as measurements and standards have become the international language of commerce and trade.

REFERENCES

[1]   *Guide to the Expression of Uncertainty in Measurement,* ISBN 92-67-10188-9, 1st
      Ed. ISO, Geneva, Switzerland, (1993): see also Taylor, B.N. and Kuyatt, C.E.,
      "Guidelines for Evaluating and Expressing the Uncertainty of NIST Measurement
      Results," NIST Technical Note 1297, U.S. Government Printing Office, Washington;
      DC, (1994).
[2]   Taylor, B.N., "Guide for the Use of the International System of Units (SI)," NIST
      Special Publication 811, 1995 Ed., (April 1994).
[3]   Sharpless, K.E., Schiller, S.B., Margolis, S.A., Brown Thomas, J., Iyengar, G.V.,
      Colbert, J.C., Gills, T.E., Wise, S.A., Tanner, J.T., and Wolf, W.R., "Certification of
      Nutrients in Standard Reference Material 1846: Infant Formula," J. AOAC Intl. 80,
      pp. 611-621, (1997).
[4]   Definitions of Terms and Modes Used at NIST for Value-Assignment of Reference
      Materials for Chemical Measurements, SP 260-136, May, W.E., Gills, T.E., et al., in
      press.

Intended Use of the IAEA Reference Materials Part I: Examples on Reference Materials for the Determination of Radionuclides or Trace Elements

A. Fajgelj, Z. Radecki, K. I. Burns, J. Moreno Bermudez, P. P. De Regge, P. R. Danesi, R. Bojanowski*

INTERNATIONAL ATOMIC ENERGY AGENCY, AGENCY'S LABORATORIES, A-2444 SEIBERSDORF, AUSTRIA

J. LaRosa

INTERNATIONAL ATOMIC ENERGY AGENCY, MARINE ENVIRONMENT LABORATORY, MC 98012, MONACO

1 INTRODUCTION

For over forty years the International Atomic Energy Agency (IAEA), through its Analytical Quality Control Services (AQCS), has prepared and distributed reference materials (RMs) to the analytical laboratories in its Member States to assist them in maintaining/improving the quality of their analytical measurements and to help them achieve internationally acceptable levels of quality assurance[1]. These RMs include matrices of environmental and biological origin, characterised for primordial and anthropogenic radionuclides, stable isotopes, organic contaminants and inorganic elements. The AQCS catalogue[2] containing a complete list of these RMs is published biannually by the IAEA. Over ninety RMs were listed in the 1998/99 catalogue which placed the IAEA among the major producers of environmental and biological matrix RMs in the world.

Preparation, characterisation and distribution of IAEA RMs is carried out by different Sections and Units of the IAEA Laboratories located in Austria and Monaco. RMs of terrestrial origin, characterised for inorganic elements and radionuclides, are mainly prepared by the Chemistry Unit (CU) located in Seibersdorf, Austria. RMs characterised for stable isotopes, H-3 and C-14 are prepared by the Isotope Hydrology Unit located in Vienna. RMs of marine origin, characterised for radionuclides, inorganic elements and organic contaminants, are prepared by Marine Environment Laboratory located in Monaco. The selection and characterisation of RMs is performed in close co-operation with other Sections in the IAEA including: Nutrition and Health Related

---

* Permanent address: Institute of Oceanology, PAS, 81-712 Sopot, Poland

Environmental Studies, Industrial Applications and Chemistry, Safeguards Analytical Laboratory and the Agriculture and Biotechnology Laboratory. For this paper only the RMs of terrestrial origin, characterised for radionuclides or inorganic elements, prepared mainly by the CU are discussed.

1.1   Common Characteristics of the IAEA-CU Reference Materials

1.1.1 Preparation and characterisation of materials.   All RMs produced thus far by the CU were prepared for and characterised as a result of IAEA intercomparison exercises. These environmental and biological RMs were collected from different places around the world. They were, with a few exceptions, processed and characterised for their physical (e.g. particle size distribution) and chemical properties (e.g. matrix composition) at the IAEA Laboratories Seibersdorf, to ensure the bulk material was of suitable homogeneity for bottling. A more detailed description of these procedures can be found elsewhere.[3-5] After bottling, the within and between bottle homogeneity tests were performed and, when satisfactory, an information sheet was prepared. The samples were then distributed to all laboratories which had expressed an interest in participating in the intercomparison exercise. The laboratories were supplied with the intercomparison sample(s) together (in some cases) with a quality control sample, the information sheet(s), the reporting forms and a diskette for reporting the results. The information sheet provided basic data about the chemical composition of the matrix, the minimum sample mass to be taken for analysis, the procedure for moisture determination, the number of samples to be analysed (6 recommended) and a list of elements or radionuclides to be determined. The participants were free to use whichever analytical technique and procedure they thought appropriate to analyse the elements or radionuclides requested. Recently (1997), in order to improve its assessment of the quality of reported results, the IAEA prepared a questionnaire in which it requested detailed information on the technique(s) used, the quality assurance and quality control procedures applied, and on the performance of the participating laboratories. It supplied this questionnaire in hard copy and on diskette together with the samples and other related documents.

1.1.2 Evaluation of intercomparison results. Intercomparison results reported to the IAEA are checked by IAEA staff and hard copies of data input files are sent back to the participants for confirmation and correction where necessary. All results for each analyte are then statistically evaluated according to the protocol outlined in Figure 1 in order to determine the distribution of the reported values. After the statistical identification and rejection of "outliers", the mean value (mean of laboratory means), the standard deviation and the limits of the 95 % confidence interval are calculated. Thus far two statistical methods have been used to identify outliers. The first method is based on a non-parametric distribution of the data and elimination of outliers by the application of tests based on Tschebytcheff's inequality[6]. The second method identifies and rejects outliers by the serial application of four statistical tests: Dixon, Grubbs, the coefficient of skewness and the coefficient of kurtosis. Results are rejected if they fail at least one of the tests. The complete procedure is repeated until no other outliers are found[7]. Recently the IAEA has employed additional criteria including "expert judgement" to evaluate results. These additional criteria are discussed below.

Preliminary Statistical Evaluation

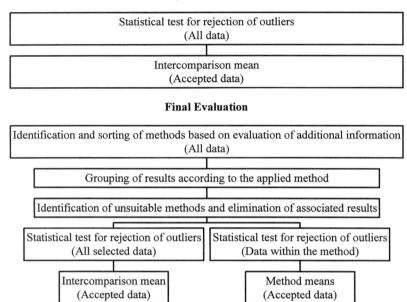

Figure 1 *Evaluation procedure of the results from the IAEA-326/327 intercomparison study including "expert judgement".*

1.1.3 Assignment of recommended and information values.[*] Regardless of the evaluation procedure, assignment of the final value as a recommended, information or listed value has depended on whether the value met certain criteria. These criteria included: a) a minimum number of laboratory means accepted for calculation of the overall mean, b) a maximum value for the relative uncertainty in the concentration or activity concentration range, c) a maximum value for the fraction of laboratory means rejected as outliers. These criteria have been summarised in Table 1 for the IAEA-326/327 intercomparison study.

There is also a set of criteria for assigning the information values, which are applied when the criteria summarised in Table 1 are only partially met. If neither of these criteria are fulfilled then the value is not assigned at all. The application of "expert judgement" to reject results not statistically detected as outliers by the methods mentioned in 1.1.2, was based on the quality of the data reported and the details of the analytical methods provided in the questionnaire by participants. The remaining data after the expert judgement were then statistically evaluated. Results below the limit of detection (if it was reported) were not considered in the statistical evaluation.

---

[*] Terms like certification, certified value, certified reference material, certificate, etc. have not been used in this paper as the majority of IAEA RMs do not strictly fulfil all the criteria to be classified as certified.

Table 1. *Criteria to be Fulfilled for Assignment of Recommended Values in the IAEA-326/327 Intercomparison Study*

	Criterion	
Parameter	*For trace elements*	*For radionuclides*
Overall mean based on	≥ 10 laboratory means and ≥ 2 different analytical methods	≥ 10 laboratory means and ≥ 2 different analytical methods
		≥ 20 laboratory means when only 1 analytical method used
Percentage of outlying laboratory means	$< 20\%$	$< 30\%$

	mg/kg	RSD	Bq/kg	RSD
Relative standard deviation for a given mass fraction or activity concentration range	> 500	$< \pm 5\%$	> 100	$< \pm 20\%$
	$100 - 500$	$< \pm 10\%$	$1 - 100$	$< \pm 30\%$
	$10 - 100$	$< \pm 20\%$	< 1	$< \pm 40\%$
	$0.1 - 10$	$< \pm 30\%$		
	< 0.1	$< \pm 40\%$		

1.1.4 Information provided to the participants in the intercomparison study. For every intercomparison, each laboratory was assigned a unique code known only to itself and the IAEA. All results submitted were reported using the laboratory code to ensure anonymity. The statistical evaluation was also included in the report where outlying laboratory mean values were identified as shown in Table 2.

Table 2. *Example of a summary of results compiled for Sr-90 in the IAEA-326 intercomparison study*

For a complete data set

Radionuclide determined	Sr-90
Number of reported laboratory means	64
Number of reported independent determinations	194
Number of accepted laboratory means	51
Number of independent determinations	163
Range of all laboratory means	0.1 - 146.7 Bq/kg
Range of accepted laboratory means	4.3 - 13.8 Bq/kg
% of outlying laboratory means	20 %
Overall mean of accepted laboratory means	10.12 Bq/kg
Standard deviation of the overall mean	± 2.1
Relative standard deviation of the overall mean	± 20.8 %
Limits95 % confidence interval	9.5 - 10.71 Bq/kg

For individual laboratory

Laboratory code	NN
Method code	B2 (Beta counting - liquid scintillation)
Number of determinations	5
Laboratory mean	19.5 Bq/Kg*
	(* identified as outlyer)
Standard deviation	± 2.8
Relative standard deviation	± 14.1 %
Limit of detection	not reported

The output file of the statistical evaluation could be presented graphically, grouped according to analysis methods or according to any other appropriate parameter.

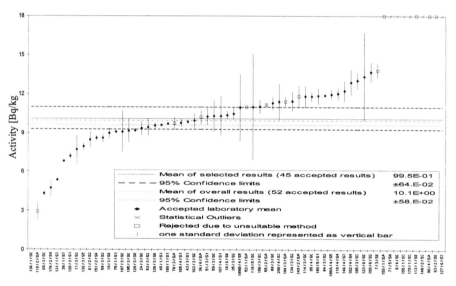

Laboratory-code / No. of determinations / Method-code

Figure 2. *Graphical representation of ^{90}Sr results in the IAEA-326 intercomparison study which included both expert judgement statistical evaluation.*

The reference sheet has included data for all analytes where the criteria for assigning recommended or information values were fulfilled. This sheet listed: recommended and the information values, the limits of the 95 % confidence interval and the number of laboratory means accepted for each analyte. Additional information provided about the RM included: designation (aim or intended use of the RM), description of the RM, information about its homogeneity, instructions for moisture determination and summary information on how the assigned values were established with a reference to the relevant intercomparison report. In certain cases, special information concerning the sample was also identified in the reference sheet, e.g. the possible presence of "hot particles" in materials collected at the areas affected by the radioactive fallout (post Chernobyl samples, samples collected in the vicinity of nuclear testing sites, etc.)

2. INTENDED USE OF REFERENCE MATERIALS

2.1 Information Provided by the IAEA on the intended use of it RMs

Before selecting any RM(s) to evaluate a chemical procedure, the user must have first defined which parameters are to be tested. To select the most appropriate or "fit to purpose" RM, the user should consult the IAEA RM reference sheet and the related report on the intercomparison study.

Information on the appropriate use of the IAEA RM is provided in the IAEA reference sheet headed "Designation". For example, the following statement appeared in the reference sheet for RM IAEA-373, Radionuclides in Grass: "This material is intended as a reference material for the measurements of radionuclides in environmental grass samples. It can also be used for the assessment and control of the laboratory's analytical work, and for validation of analytical methods used in a laboratory, i.e., in general quality assurance within a laboratory and for training purposes."[8,9] This information provided by the IAEA is of a general nature. However, there exist many international standards and guides where the appropriate use of reference materials has been described. In general, for IAEA RMs prepared more than ten years ago the information on the intended use was not included in the relevant reference sheet, e.g. IAEA Soil-6, IAEA/V-10 Hay Powder, etc.

2.2   International Guidelines Regarding Intended Use of RMs and Applicability to the IAEA RMs

The following sections summarise international guidelines on RMs and the present status of IAEA RMs with respect to meeting each guideline. For each section it is assumed that a RM was selected, which mimics the sample as close as possible with respect to chemical composition and concentration levels of the elements/radionuclides of interest.

2.2.1 Reference materials and assigned property values[10]. The "property values" assigned to the IAEA RMs discussed in this paper are the mass fraction of trace/minor elements or the activity concentration in the case of radionuclides. Once the "property values" have been assigned, an inherent attribute of a RM is that its property value must remain stable over a defined period of time (expiration date defined). This must be supported by stability tests and ensured by maintaining it under proper storage conditions. The IAEA RMs have been selected and characterised for those analytes which are stable in storage under ambient conditions encountered in routine laboratory conditions. To confirm the stability of the IAEA RMs, stability tests must be carried out at regular intervals (at least once per year). The stability of IAEA RMs has been demonstrated many times (see section 2.2.7) and therefor the IAEA RMs fulfil the criterion to ensure stability of the RM "property values".

2.2.2 Establishing traceability of a measurement result[10]. Traceability as it is used in this report is defined according to the International Vocabulary of Basic and General Terms in Metrology (VIM): "Traceability is the property of the result of a measurement or the value of a standard whereby it can be related to a stated reference, usually national or international standards, through an unbroken chain of comparisons all having stated uncertainties."[11] For IAEA RMs of terrestrial origin characterised for radionuclides or inorganic elements, the assigned values are expressed either as an activity concentration (Bq/kg, etc.) or element concentration (in some cases as a mass fraction) respectively. The kilogram and becquerel are International System of Units (SI) or units derived from the SI and according to the definition above, the property values of an RM must be traceable to them. However, this criterion is not fulfilled when the assignment of the RM property value is based on the results of an intercomparison. Although single results or a group of results might be traceable to SI, the statistical evaluation applied to calculate the mean value destroys the traceability chain. At the IAEA Consultants' Meeting on "Traceability of IAEA-AQCS Reference Materials to SI-Units"[12], it was clearly pointed out that the assigned property values of current IAEA RMs could be considered as traceable to the respective laboratory intercomparison only and not to any other further point of reference.

The traceability issue with respect to natural matrix RMs has to be viewed in a wider perspective than within the IAEA. Most of the internationally available matrix RMs characterised for trace elements or radionuclides originate as a result of intercomparison studies and therefore fail the traceability criterion. A report, published recently by Papadakis et al., describes the establishment of a SI-traceable copper concentration in the Antarctic Sediment candidate RM by using isotope dilution combined with inductively coupled plasma mass spectrometry.[13] This is an example where a primary method of measurement was applied. However, when these results are combined with other results derived from comparative methods of measurement (i.e. not primary) obtained in the certification campaign, the traceability of the assigned value becomes dubious (questionable?).

Although the ISO Guide 33 considers the establishment of traceability by the use of RMs; in the case of chemical RMs, only pure substances, their solutions, alloys, and gas mixtures would currently fulfil the requirements for traceability (see also 2.2.4), while the natural matrix RMs do not. It is important not to confuse the terms "traceability" and "accuracy" of the assigned property value. Even when the assigned value of an intercomparison study, is considered a "the best estimate of the property or true value", this does not by itself assure that the result is traceable. Considerable work still remains to be done in this field, including the production of traceable matrix RMs, guidance on their use and definition of their role in the analytical process once they become available. From the discussion above it must be concluded that the current IAEA RMs can not be used to establish traceability of a measurement result.

2.2.3 Determining the uncertainty of the measurement results.[10, 14] The uncertainty of a measurement result is defined in VIM as a: "Parameter, associated with the result of a measurement, that characterises the dispersion of the values that could reasonably be attributed to the measurand."[11] (In case of the IAEA RMs measurand is a mass fraction of trace elements or activity concentration of radionuclides respectively.) Here the term "uncertainty" refers to the expanded combined uncertainty, which takes into account all sources of uncertainty associated with the relevant chemical measurement process. As noted in section 1.1.1, reporting of the uncertainty of measurement results is routinely requested from the participants in IAEA intercomparison exercises. However, in the majority of cases uncertainties are not reported and the few that are would only account for some of the sources of uncertainty e.g. uncertainty due to the counting statistics in γ-spectrometric measurements. It must be pointed out that the statistical evaluation employed by the IAEA does not take into account the measurement uncertainty reported by the participants. All calculations are performed using only the accepted laboratories' mean values.

As is the case with traceability, the uncertainty in the assigned values obtained through an intercomparison study is still an open question. J. Pauwels et al. have proposed a pragmatic method for the determination of the uncertainty of property values derived from an interlaboratory comparison.[15] Even such a pragmatic method requires the establishment of a full uncertainty budget for each laboratory result. Quantification of uncertainty in chemical measurements is a developing issue. With the issuance of the EURACHEM Guide on Quantifying Uncertainty in Analytical Measurements[14], the basic principles and examples were provided to the analytical chemistry community, but time is needed for those principles to be fully appreciated and implemented in all laboratories. For this reason, characterisation through a small number of selected laboratories fulfilling the highest quality criteria and employing primary methods for measurements is the preferable approach for future certification of the IAEA RMs.

Although the use of RMs for uncertainty quantification is discussed in both ISO Guide 33[10] and the EURACHEM Guide[14], neither provides guidance on how this should be done. A possible option is to use a RM for quantification of a specific source of uncertainty, e.g. sample digestion, recovery, etc. In this case a sufficient number of independent experiments under the same experimental conditions has to be performed. The standard deviation of the results may be ascribed as a standard uncertainty (however the distribution of data must be considered) and then included in the calculation of the combined uncertainty. It must be pointed out that for this purpose to apply the RM must be homogeneous with respect to the element/radionuclide measured From the results of independent measurements the laboratory can establish its own mean value for a specific investigation and calculate its standard deviation for obtaining standard uncertainty. Examples of such an evaluation can be: quantification of spectral interferences in γ-spectrometry, quantification of background and matrix effects in neutron activation analysis, etc. When the homogeneity of the material satisfies the requirements of the user, the IAEA RMs can be used for evaluation of uncertainty sources - quantification of standard uncertainties.

2.2.4 Calibration of an apparatus.[10] Quality control/assurance procedures require all measuring instruments, glassware, balances, etc., to be calibrated against traceable standards at regular intervals.[16, 17] Especially important is the calibration of instruments (different spectrometers, etc.) that provide a result directly linked to measurands such as amount of substance, concentration, activity, activity concentration, etc. These results are normally obtained from a calibration curve or by relative measurements. In such cases, the calibration is usually performed using a calibration standard prepared from a pure chemical substance, radioactive isotope or a mixture - multielement calibration standard. These calibration standards are often available as certified reference materials.* They are characterised by primary methods of measurements, i.e., gravimetry, titrimetry, coulometry, etc. and are directly traceable to the mole (SI unit for the amount of substance). As discussed in 2.2.2, in the case of natural matrix RMs characterised by an intercomparison study, traceability to SI units is not established. For this reason most of

---

*Terminology in respect to calibration standards, certified reference materials, primary standards, secondary standards, etc. is not fully supported by quality requirements and classification of these materials. Further systematisation is required at the international level.

the currently available natural matrix RMs, including those produced by the IAEA can not be used for the instrument calibration. (Special cases, like laser ablation techniques, where no other possibility for calibration than a use of matrix RM exists, require additional consideration.)

2.2.5 Assessment of a measurement method.[10] One of the basic principles of quality assurance in analytical chemistry is that only validated methods can be used for routine analyses. Even in cases when the methods are available from the literature and peer-verified methods (intercomparison organised with various laboratories) they must be validated in the particular laboratory that is going to apply them. Methods developed in-house should undergo a complete method validation, which includes: accuracy (trueness + precision), sensitivity, selectivity, linearity, limit of detection, possible interferences, recovery and ruggedness. The use of a RM containing the analyte at a known concentration is one of the most appropriate tests to validate the method and the performance of the analyst.[18] For assessment/validation of accuracy, precision and possible interferences, the composition of RM should approximate that of the actual samples and should contain the analyte at about the same concentration. For validation of sensitivity, selectivity, linearity, recovery and ruggedness, tests should be carried out at different concentration levels of the analyte. When a limit of detection is assessed, then a RM which does not contain the analyte of interest, but is similar to the actual sample in respect to matrix, would be needed. For method validation normally more than one RM or a selected RM spiked with the analyte of interest at different concentrations levels is used. When selecting RMs for method validation purposes, one finds that the availability of natural matrix RMs is limited. For example, natural matrix RMs certified for radionuclides are available only from a few producers and the majority of these RMs were prepared by the IAEA. This is illustrated in Table 1 for RMs with an assigned property value for ^{90}Sr.[19]

Table 1 *Available Environmental and Biological Reference Materials for Determination of ^{90}Sr.*

Name	Code	Status	Concentration (Bq/kg)
Mediterranean Tuna Fish	IAEA-352	N	0.2
Sea Plant	IAEA-307	N	0.72
Pacific Ocean Water	IAEA-298	C	1.32
Milk Powder	IAEA-A-14	C	1.5
Marine Sediment	IAEA-368	N	1.8
Milk Powder	IAEA-321	C	3.3
Cockle Flesh	IAEA-134	N	4.8
Columbia River Sediment	NIST-SRM 4350B	N	5.3
Radionuclides in Whey Powder	IAEA-154	C	6.9
Rocky Flats Soil	NIST-SRM 4353	C	7.63
Milk Powder	IAEA-152	C	7.7
Clover	IAEA-156	C	14.8
River Sediment	GBW 08304	C	19.7
Peruvian Soil	NIST-SRM 4355	N	22
Soil	IAEA-SOIL-6	C	30.34

Name	Code	Status	Concentration (Bq/kg)
Animal Bone	IAEA-A-12	C	54.8
Marine Sediment	IAEA-135	N	64.5
Marine Sediment	IAEA-367	C	102
Soil	IAEA-375	C	108
Freshwater Lake Sediment	NIST-SRM 4354	C	1090
Grass	IAEA-373	C	1320

Note: N = non-certified value   *C = certified (NIST, GBW)*
recommended value (IAEA)

When selecting an IAEA RM for method validation or performance assessment of the analyst, the user has to be aware that the results obtained by analysing these RMs can only be compared with the statistically derived mean of the accepted results of the participating laboratories. For comparing a laboratory mean value with the overall mean value of the intercomparison study (reference or information value) the mathematical formulae from ISO Guide 33 for assessment of trueness (the closeness of agreement between the average value obtained from a series of test results and an accepted reference value) can be followed:

$$-a_2 - 2\sigma_D \leq \bar{x} - \mu \leq a_1 + 2\sigma_D \qquad (1)$$

where:
$\bar{x}$ is the RM user's mean value
μ is the recommended or information value (mean of laboratory means)
$a_{1,2}$ additional level of uncertainty defined by the user to meet the economic or technical limitations or stipulation
σ_D is the standard deviation associated with the RM user's measurement process. It is combined from the within-laboratory fluctuation (σ_w given as standard deviation s_w - RM user's results) and the between-laboratories fluctuation (σ_{Lm} is a standard deviation of statistically accepted intercomparison means - information available from intercomparison report), as follows:

$$\sigma_D^2 = \sigma_{Lm}^2 + \frac{s_w^2}{n} \qquad (2)$$

where:
n   is the number of RM user's independent determinations

The information on between-laboratories standard deviation (σ_{Lm}) is given in the related IAEA report on the intercomparison study, while the reference sheet will normally provide the reference value, the limits of the 95 % confidence interval and number of accepted laboratory means. It would be wrong to expect that only the user's results that fall between the limits of the 95 % confidence interval are satisfactory in comparison with the recommended value. The 95 % confidence interval is a statistically derived value and depends on the number of accepted laboratory means (p) in an intercomparison study. In some of the latest IAEA intercomparison studies the number of accepted means was larger than 50 and for this reason the confidence interval became unreasonably small. The value for the confidence interval should not be considered independently from other parameters (p, σ_{Lm}), refer to Figure 2 for an illustration. As already pointed out, the user of the RM should consult the intercomparison report before using the IAEA RMs and drawing any conclusion as to the accuracy of his own results.

When the precision of the method is assessed by application of a RM, the user should compare the standard deviation of his results (s_w, within-laboratory standard deviation) with the required value of the within-laboratory standard deviation (σ_{wo}). The required within-laboratory standard deviation (σ_{wo}) may be set out by the user himself as a target value or defined by the customer of the analytical results as a specific requirement for the acceptance of these analytical results. The ratio between s_w and σ_{wo} has to be calculated:

$$\chi_c^2 = \left(\frac{s_w}{\sigma_{wo}}\right)^2 \qquad (3)$$

and χ_c^2 compared with the tabulated value of χ_{Table}^2, where $\chi_{Table}^2 = \frac{\chi_{(n-1);0.95}^2}{n-1}$.

The method assessed will be regarded as precise as required when $\chi_c^2 \leq \chi^2$.

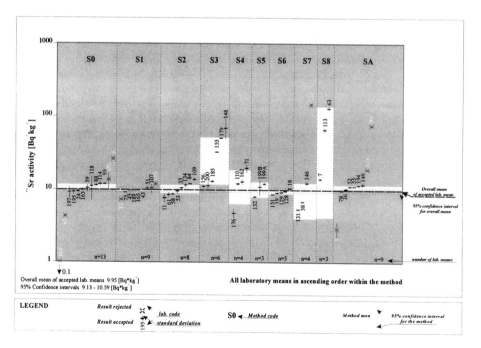

Figure 3 *IAEA-326 intercomparison results grouped according to the methods applied.*

2.2.6   Use of reference materials for recovery studies.[20]   In principle, when no better possibility exists (use of radioactive tracers, isotope dilution techniques, etc.), recoveries could be estimated by the analysis of RMs. The recovery is the ratio of the concentration of analyte found to that stated to be present. Results obtained on test materials of the same matrix could, in principle, be corrected for recovery on the basis of the recovery found for the RM. However, the RM user has to be aware of several problems which potentially beset this use, namely: a) the validity of any such recovery estimate depends on the premise that the analytical method is otherwise unbiased; b) the range of appropriate matrix RMs available is limited; and c) there may be a matrix mismatch between the test material and the most appropriate RM available.

In the last instance the recovery value obtained from the RM would not be strictly applicable to the test material.   The shortfall applies especially in sectors such as foodstuffs analysis where RMs have to be finely powdered and dried to ensure homogeneity and stability.   Such treatment is likely to affect the recovery in comparison with that pertaining to fresh foods of the same kind.   In case of trace element or radionuclide determinations in environmental and biological samples, i.e., soil, vegetation, human tissue, etc. the preparation of samples will be very similar to the preparation of RMs and also a digestion procedure will often be applied in the analytical process. In such a case a RM is a proper tool to assess the recovery or chemical yield. The IAEA RMs can be used for this purpose.

2.2.7   Use of reference materials for quality control purposes.   Quality control practice requires use of quality control materials, duplicate and blank analysis and continuous observation of the analytical performance of the measurement system, including the operator. This is normally supported by the use of statistical analysis and control charts. It is not always possible to prepare a suitable quality control material in the laboratory and quite often the RMs are used for this purpose. In Figure 4 a control chart is presented showing the performance of a laboratory in $^{239 + 240}$Pu analysis using RM IAEA Soil-6 as a quality control material. This laboratory also took part in the intercomparison study and was included in the statistical evaluation of data, having its mean value very close to the recommended value for $^{239 + 240}$Pu. Some trends evident from the control chart in decreasing values, especially from 1993 to 1994, have been thoroughly studied. The reason was found to be a change in concentration of the calibration standard solution due to evaporation. After a new solution was prepared, the results were again closer to the assigned mean value. As it is evident from the control chart, the majority of results are within the limits of the 95 % confidence interval over a period of seven years.

2.2.8   Other applications.   The IAEA RMs have been successfully used for assessment of the performance of the IAEA network laboratories, e.g. the network of Analytical Laboratories Measuring Environmental Radioactivity (ALMERA). Although the mean value of such a specific laboratory intercomparison with a limited number of selected laboratories might be slightly different from the mean value of the original world-wide intercomparison, a consensus value can be established and the relative performance between the laboratories assessed accordingly. Such an intercomparison provides a sound basis for establishing the comparability of the results obtained in different laboratories analysing the same type of samples. However, there is a limitation for such type of application as the RMs are normally available to the users in limited quantities.

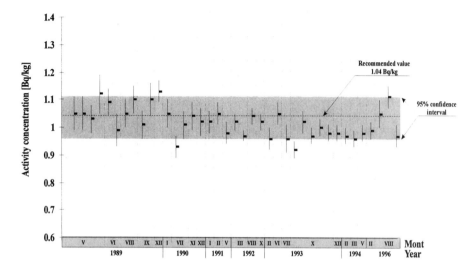

Figure 4 $^{239+240}Pu$ *activity concentration measured in the IAEA-Soil-6 RM over the period 1989 - 1996.*

3 FUTURE PLANS

Aware of the limitations of its RMs as outlined in this paper, the IAEA is following the latest metrological developments and requirements in the area of analytical chemistry. The IAEA efforts in the organisation of world-wide intercomparison studies continues to be regarded by its Member States as a valuable contribution to the laboratories. The IAEA intends to continue to provide this assistance, however, in a modified manner consistent with the new requirements associated with the use of RMs.

The interest in the IAEA RMs is high and around 1000 units are sold per year as shown in Figure 5. For this reason the IAEA plans to upgrade some of its RMs so that traceability either to the basic SI unit the mole (amount of substance) or derived unit Bq (activity) can be clearly established. Materials have already been selected for upgrade based on availability and user demand. Table 3 lists the RMs under consideration for upgrading involving the determination of radionuclides.

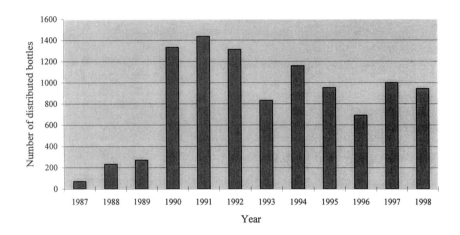

Figure 5   *Number of the IAEA RMs units sold per year. Figure for 1998 includes sales
up to November (Figure 5 does not include RMs produced and distributed by
the Hydrology Unit neither units which were given free of charge in support of
the IAEA programmes.)*

    For the determination of trace elements in biological or environmental samples, a
similar approach is planned. A selection of a limited number of expert laboratories,
performing the analysis according to prescribed protocols and with a quantified
measurement uncertainty, should produce results for the certification of the IAEA RMs.
Two sets of materials are already under investigation: the IAEA-390, a set of three algae
materials, and a set of two membrane filters artificially loaded with urban dust collected in
Vienna and in Prague.

Table 3   *IAEA Intercomparison Materials Selected for Upgrade.*

Code	Type	Analytes of Interest
IAEA-152	Milk Powder	K-40, Sr-90, Cs-137
IAEA-312	Soil	Ra, Th and U
IAEA-314	Stream Sediment	Ra, Th and U
Soil-6	Soil	Sr-90, Cs-137, Ra-226 and Pu-239
SL-2	Lake Sediment	K-40, Sr-90, Cs-137, Pb-210, Ra-226, Ra-228, Th-228, Th-234, U-238 and Pu-239/240
IAEA-375	Soil	K-40, Sr-90, Ru-106, Sb-125, I-129, Cs-134, Cs-137, Th-232 Ra-226, Th-228, U-234, Pu-238, U-238, Pu239/240 and Am-241

4 CONCLUSION

With ever increasing and stricter requirements being placed on the quality of analytical results by customers and regulators, the need for appropriate RMs is continuously growing. Metrological concepts pertaining to analytical chemistry require traceability of analytical measurements to be established and measurement uncertainty to be reported together with the test results. However, there is still a discrepancy between the current requirements and the tools available to fulfil these requirements. Natural matrix RMs that already fulfil the criteria for traceability to the SI units are very rare because the producers of RMs are facing the same problem as the analytical community: a) lack of primary methods of measurement, b) procedures for uncertainty quantification in analytical measurements are still not fully established or completely harmonised, c) most of the international standards related to traceability and measurement uncertainty are based on the "nature" of physical measurements, which can not directly be transferred to chemical measurements. In addition, in the absence of primary methods of measurements, it is necessary to apply more than one analytical technique for the characterisation of RMs. The internationally accepted guidance for reporting the uncertainty of such combined results is still under development. Also needed is a better defined hierarchy of RMs according to the fulfilled metrological requirements and the definition of the role of the RMs in the measurement process. RMs of different quality will always be available and a proper terminology should be developed to distinguish between them. What is the role of SI traceable RM once when available? There is no doubt that such materials would transfer the property value which will be linked to the most proper point of reference (basic SI unit(s)) and comparability of the results will be possible. But, would such a material then be used for calibration, correction of results, etc., or for quality assurance purposes as it is now? Would a traceable RM be used to link the test results on similar type of sample to the SI units? Even when these questions will be clearly answered, the RMs producers will still have the same objective: assuring that the values assigned to the RMs are the best estimates of the true value. However, there should be clear evidence how this was achieved and what is the meaning of the assigned property values and their uncertainties.

As described in this paper, the IAEA intends to follow recent developments in quality requirements for RMs. As traceable natural matrix RMs are not expected to be available in the near future, many of the RMs that are still in stock will continue to be used. It remains the responsibility of the RM user to obtain the necessary information on the quality of the assigned property values for the RM, its intended use and to use these RMs properly.

References

1.   R.M. Parr, A. Fajgelj, R. Dekner, H. Vera Ruiz, F.P. Carvalho, P.P. Povinec, *Fresenius J Anal Chem*, 1998, **360**, 287.
2.   International Atomic Energy Agency, 'IAEA AQCS Catalogue for Reference Materials and Intercomparison Exercises 1998/1999', Vienna, Austria, 1998.
3.   V. Strachnov, V. Valkovic, J. LaRosa, R. Dekner, R. Zeisler, *Fresenius J Anal Chem*, 1993, **345**, 169.
4.   R. Zeisler, R. Dekner, V. Strachnov, H. Vera Ruiz, *Fresenius J Anal Chem*, 1995, **352**, 14.

5. A. Fajgelj, R.M. Parr, R. Dekner, P.R. Danesi, V. Valkovic, H. Vera Ruiz, The IAEA Analytical Quality Control Services (AQCS) Programme on Intercomparison Runs and Reference Materials, Proceedings of an International Symposium on Harmonisation of Health Related Environmental Measurements Using Nuclear and Isotopic Techniques, Hyderabad, India, 4-7 November 1996, International Atomic Energy Agency, Vienna, 1997, p. 175.
6. L. Pszonicki, *Anal. Chim. Acta*, 1985, **176**, 213.
7. R. Dybczynski, *Anal. Chim. Acta*, 1980, **117**, 53.
8. V. Strachnov, J. LaRosa, R. Dekner, A. Fajgelj, R. Zeisler, Report on the Intercomparison Run IAEA-373: Determination of Radionuclides in Grass Sample IAEA-373, IAEA/AL/073, International Atomic Energy Agency, Vienna, 1996.
9. Reference Sheet IAEA-373, Radionuclides in Grass, International Atomic Energy Agency, Vienna, Austria, 1994.
10. ISO Guide 33, 'Uses of Certified Reference Materials', Revision of ISO Guide 33: 1989, International Organisation for Standardisation, Geneva, Switzerland, 1998, to be published.
11. 'International Vocabulary of Basic and General Terms in Metrology', International Organisation for Standardisation, Geneva, Switzerland, 1993.
12. Report on the Consultants' Meeting on Traceability of IAEA-AQCS Reference Materials to SI-Units, IAEA/AL/105, International Atomic Energy Agency, Vienna, Austria, 1996.
13. I. Papadakis, P. D. P. Taylor, P. De Bievre, *J. Anal. At. Spectrom.*, 1997, **12**, 791.
14. 'Quantifying Uncertainty in Analytical Measurements', EURACHEM, First edition, 1995,
15. J. Pauwels, A. Lamberty, H. Schimmel, *Accred Qual Assur* , 1998, **3**, 180.
16. F. E. Prichard, N. T. Crosby, J. A. Day, W. A. Hardcastle, D. G. Holcombe, R.D. Treble, 'Quality in Analytical Chemistry Laboratory', Editor E.J. Newman, John Wiley & Sons Ltd. Chichester, England, 1995.
17. ISO/IEC Guide 25 : 1990 (E), 'General Requirements for the Competence of Calibration and Testing Laboratories', International Organisation for Standardisation, Geneva, Switzerland, 1990.
18. European Commission, BCR information, 'Metrology in Chemistry and Biology: A Practical Approach', Report EUR 18405 EN, Luxembourg, 1998.
19. 'Survey of Reference Materials', Volume 1, Biological and environmental reference materials for trace elements, nuclides and microcontaminants, IAEA-TECDOC-854, International Atomic Energy Agency, Vienna, 1995.
20. M. Thompson, S. L. R. Ellison, A. Fajgelj, P. Willetts, R. Wood, 'Harmonised Guidelines for the Use of Recovery Information in Analytical Measurement', IUPAC Technical Report, 1998, to be published in Pure and Applied Chemistry.

Intended use of the IAEA Reference Materials Part II: Examples on Reference Materials for Stable Isotope Composition

M. Gröning[1,2] K. Fröhlich

[1] INTERNATIONAL ATOMIC ENERGY AGENCY, ISOTOPE HYDROLOGY SECTION, A-1400 VIENNA, AUSTRIA

P. P. De Regge, P. R. Danesi

[2] INTERNATIONAL ATOMIC ENERGY AGENCY, AGENCY'S LABORATORIES, A-2444 SEIBERSDORF, AUSTRIA

1 INTRODUCTION

In this paper IAEA reference materials distributed by the Isotope Hydrology Unit will be discussed.[1] These reference materials are intended for the determination of the ratios of stable isotopes of light elements such as hydrogen, carbon, nitrogen, oxygen and sulfur in environmental materials. Some of the available materials are primary reference materials and define conventional scales for reporting of measurement results.

An overview on other IAEA reference materials and the general scope of the Analytical Quality Control Services (AQCS) is given in an accompanying paper[2] as well as in the bi-annually published AQCS catalogue.[3]

The Isotope Hydrology Unit is also responsible for the distribution of ^{14}C quality assurance materials. These materials are intended for laboratories performing ^{14}C activity measurements on natural materials and at environmental activity levels. A detailed description of these materials and their characteristics can be found elsewhere.[4-5]

1.1 Common Characteristics of the Stable Isotope Reference Materials

The reference materials distributed by the Isotope Hydrology Unit are chemically pure compounds such as carbonates, sulfates, sulfides, nitrates, graphite or polyethylene, or nearly pure natural materials like distilled water, carbonate rock, silicates, refined oil, sugar, cellulose and similar compounds.[1] These materials are distinct from the other IAEA environmental matrix materials.

Most of these materials have been prepared to serve as reference materials. Therefore much care was taken in the initial purification and homogenisation of the raw material. The majority of all 30 stable isotope reference materials was produced by scientists under IAEA technical contracts or by cooperating institutions like the National Institute for Standards and Technology (NIST) in USA or the United States Geological Survey (USGS). Some of the materials were produced directly by the IAEA Isotope Hydrology Unit. In most cases the recommended values for stable isotope ratios in these materials were determined by interlaboratory comparison exercises. For those materials issued already twenty or more years ago, the number of participating laboratories was rather limited (in some cases less than ten).

1.2 Terms used for different categories of reference materials in this paper

Several definitions for international distributed reference materials and for internal laboratory standards have to be clearly distinguished. Unfortunately there are no clear guidelines on the definitions to be used in this field. Some of the expressions are ambiguous, some other used in different context by different authors. For this paper the definitions used are explained below.

These definitions used for the various kind of materials should be clearly distinguished from each other:

- *Primary reference material[1] (or international standard)[6]*: a natural, synthetic or virtual material versus which, by general agreement, the relative variations of stable isotope ratios in natural compounds are expressed. It is used to define a conventional scale for reporting variations of stable isotope ratios.
- *Calibration material[1] (or primary standard)[6]*: a natural or synthetic compound which has been carefully calibrated versus the primary reference material, and whose calibration values have been internationally agreed and adopted. It is used in case the primary reference material is not existing or available to calibrate measurements and instruments. Each primary reference material can be referred to as a calibration material as well.
- *Reference material (RM)*: a natural or synthetic compound which has been carefully calibrated versus the primary reference material and whose property values are sufficiently homogeneous and well established and are associated with well determined uncertainties. It is used to calibrate laboratory equipment and measurement methods for analysis of materials of a composition different from that of the primary reference material. The available reference materials cover a broad spectrum of chemical compositions and a wide range of stable isotope ratios. Most existing stable isotope ratio reference materials were first investigated in interlaboratory comparison exercises and distributed as intercomparison materials. Since they were used de facto as reference materials to calibrate equipment and measurements, some years ago their status was adjusted accordingly.
- *Intercomparison material*: a natural or synthetic compound with proven homogeneity which provide the means to check the overall quality of measurements performed in comparison with that of other laboratories. Its isotopic composition is computed by averaging the results of several laboratories obtained in interlaboratory comparison exercises and in individual assays, after elimination of outliers using a 2σ interval criterion.[7]
- *Working standard (or transfer standard)[6]*: This term in stable isotope mass spectrometry is somewhat misleading since it is not describing a 'standard', but an arbitrarily chosen gas used as a reference for analysis of isotope ratios of samples in a dual-inlet mass spectrometer (see section 1.3). All measurements of prepared samples and reference materials are made versus this working standard and are later converted into an international accepted scale. A better term would be 'laboratory reference gas'
- *Internal standard (or reference standard)[6]*: This term describes materials which are carefully selected and of similar composition as the normal samples and which are

used routinely as a standard to calibrate or check measurements and the measuring instruments.

1.3 Conventional Scales for Reporting Stable Isotope Ratios

The analysis for stable isotope ratios is almost exclusively performed by mass spectrometry. The samples are firstly converted into a suitable gas containing isotopes of the element under investigation and are then analysed in dual-inlet mass spectrometer, where the isotopic composition of the sample gas is directly compared with that of a gas working standard. The analytical procedure includes also the measurement of internal laboratory standards of similar chemical composition as the sample. The internal laboratory standards have to be carefully calibrated versus the available reference materials and ensure the traceability of results.

The natural variations of the stable isotope ratios of light elements are rather small. For example, the total range of the variation of $^{18}O/^{16}O$ in water and ice on earth is in the order of only seven percent and increases to about ten percent for all natural oxygen bearing materials.[8] For hydrogen isotopes with their large relative mass differences the variations of $^2H/^1H$ reach higher values of about 80 % for natural hydrogen bearing compounds.

In virtually all applications of stable isotopes in earth sciences the relative deviation of the isotopic ratio from a standard is of interest rather than the "absolute" isotopic ratio of the given sample. Therefore, a material is selected as a primary standard and its stable isotope ratio defines the zero-point of a relative conventional scale. The easiest example is nitrogen, where the isotopic composition of nitrogen in atmospheric air is taken as the primary standard. For convenience the measurements are not reported as isotope ratios, but given as relative deviation from the isotope ratio of the standard according to the following formula:

$$\delta = \frac{R_{SA} - R_{ST}}{R_{ST}} \tag{1}$$

with δ (e.g. δ^2H, $\delta^{13}C$, $\delta^{15}N$, $\delta^{18}O$, $\delta^{34}S$) being the normalised difference of the isotope concentration ratios R ($^2H/^1H$, $^{13}C/^{12}C$, $^{15}N/^{14}N$, $^{18}O/^{16}O$, $^{34}S/^{32}S$) of the sample and the standard (e.g. $\delta^{15}N$ with air-N_2 as standard). As the differences between sample and standard are normally very small, the δ-values are usually expressed as per mille difference (parts per thousand).[9]

The accuracy of measurements in experienced laboratories is in the order of ±0.1 ‰ for $\delta^{18}O$ and ±1 ‰ for δ^2H or even better. This is possible due to the special analysis arrangement with direct comparison of sample isotope ratios to those of the working standard during the measurement and due to the daily calibration of the mass spectrometer and the sample preparation lines with suitable internal standards (internal laboratory sub-standards calibrated against international reference materials).

The so called δ-scales for stable isotope ratios of the elements hydrogen, carbon, nitrogen, oxygen and sulfur are good examples for conventional scales. The knowledge on the absolute isotope ratio of a given material is not necessary for their use, since these scales are defined completely arbitrarily versus a chosen primary reference material. For oxygen and hydrogen ocean water was chosen as the largest accessible reservoir on earth and in view of the slight variations in isotopic composition throughout the oceans a hypothetically well mixed mean ocean water was selected as the reference (called SMOW

- Standard Mean Ocean Water). This is a good example for a scale based on a virtual material not realised in nature. In case of carbon, marine carbonate of organic origin in a certain geological formation in the USA was selected (PeeDee Belemnite, PDB); for nitrogen, atmospheric nitrogen was the obvious choice; and for sulfur, material from a meteorite was selected and expected to be representative for the mean cosmic abundance of sulfur isotopes (Cañon Diablo Troilite, CDT).

However, the main persisting problem was the proper calibration between different laboratories since the primary standard for oxygen and hydrogen did not exist at all, and both original materials defining the δ-scales for carbon and sulfur (PDB and CDT) were exhausted since some time. The IAEA therefore initiated the production of new reference materials which were calibrated with the best available techniques relative to the original δ-scales of the respective elements. Beside many other materials, these activities resulted in the production of the three materials VSMOW (water), NBS19 (carbonate), and IAEA-S-1 (silver sulfide), which were adopted by international agreement as calibration materials after extensive tests by experienced laboratories. These calibration materials were used to define the zero-point of new δ-scales named VSMOW, VPDB and VCDT which should be as close as possible to the original SMOW, PDB and CDT δ-scales (the V standing for Vienna in all three cases).

As an example the development of the scales in the case of hydrogen and oxygen isotope ratio standards will be presented.

1.4 Definition of the Scales for Reporting Hydrogen and Oxygen Isotope Ratios

First the historical development is presented to establish the scales for reporting hydrogen and oxygen isotope ratios, primarily intended for measurements on water samples and extended to other hydrogen and oxygen bearing materials.

Already in 1953 'average ocean water' was suggested and used as reference point for measurements.[10] Since no 'average ocean water' exists, this concept was refined in 1961 by defining the hypothetical Standard Mean Ocean Water (SMOW).[11] Its isotopic composition was defined as a weighted average of the available measurements of the isotopic composition in the main oceanic water masses. But since SMOW was just a concept and never existed as real water sample, it couldn't be used for calibration of laboratory measurements.

However, the isotopic ratios of SMOW were defined with respect to the existing water standard NBS-1[12] of the US National Bureau of Standards, used earlier for an interlaboratory comparison. So for the first time a physically existing material was used to calibrate different laboratories to the SMOW scale. NBS-1 was readily available for world-wide distribution together with another water standard called NBS-1A obtained from melted snow with an lower abundance of the heavier isotopes. During an IAEA interlaboratory comparison in 1965 serious doubts were confirmed concerning the preservation of the NBS-1 water standard and possible changes of its isotopic composition.

At an IAEA Panel Meeting in 1966 it was therefore recommended to establish a pair of two new standards, the first one being as close as possible to the defined SMOW and the other one with an abundance of the heavier isotopes close to the lowest limits observed in natural water.

The new standard with an isotopic composition as close as possible to SMOW was prepared by H. Craig. It was obtained by mixing distilled ocean water with small amounts of other waters in order to adjust its isotopic composition as close as possible to that of the defined SMOW. This task was complicated due to the required adjustment of both the isotopic composition of hydrogen and of oxygen. This standard was ready in 1968 and was called Vienna Standard Mean Ocean Water (VSMOW). According to the control analyses performed by Craig VSMOW has the same $^{18}O/^{16}O$ ratio as the defined SMOW, but a slightly lower $^2H/^1H$ ratio or respectively a slightly negative δ^2H value (-0.2‰). However, this slight difference is about a factor of four to five lower than the analytical uncertainty of most laboratories. Absolute isotope ratios on VSMOW were determined for $^{18}O/^{16}O$,[13] for $^{17}O/^{16}O$ [14] as well as for $^2H/^1H$.[15,16,17]

The second standard was obtained by E. Picciotto from melting a firn sample at Plateau Station, Antarctica. This material was named Standard Light Antarctic Precipitation (SLAP). The absolute isotope ratios of SLAP were determined only for $^2H/^1H$,[15,16,17] absolute $^{18}O/^{16}O$ ratios for SLAP were computed from those of VSMOW by using its assigned value of $\delta^{18}O$=-55.5 ‰ versus VSMOW (see below).

All four materials VSMOW, SLAP, NBS-1, NBS-1A (the latter both transferred to IAEA from the National Bureau of Standards) were then distributed by the IAEA Isotope Hydrology Laboratory.

In 1976 an IAEA Consultants' Meeting was convened in order to discuss the isotope results on these standards obtained so far from laboratories and to advise on future action on standardization of stable isotope measurements.[18]

The recommendation of the experts concerning water standards was to express all future results as δ-values relative to VSMOW in order to resolve confusion on results expressed in non-corresponding scales. It was stated that the coherence between δ-values reported by different laboratories could be improved by adopting fixed δ-values for a second water reference standard. The experts recommended to adopt SLAP for this purpose and to normalize the ^{18}O and ^2H δ-scales relative to this standard. In Table 1 the δ-values are listed with those of VSMOW being by definition at zero and those of SLAP established by the weighted mean of results from different laboratories. The established δ^2H-value for SLAP is in remarkable good agreement with the available absolute hydrogen isotope ratio determination of the two materials using isotope dilution methods (deviation expressed as δ^2H smaller than 0.5‰).

Table 1   *Oxygen and hydrogen δ-values versus VSMOW assigned to the existing water reference materials (NBS-1 and NBS-1A δ-values are reported versus SMOW)*

Name	$\delta^{18}O$ [‰]	δ^2H [‰]
VSMOW	0	0
SLAP	-55.5	-428
GISP	-24.8±0.05	-189.5±1.0
NBS-1	-7.94	-47.6
NBS-1A	-24.33	-183.3

The definition of VSMOW as zero of the oxygen and hydrogen δ-scales and the adoption of fixed δ-values for SLAP is therefore a slight modification of the original definition of a δ-scale in formula (1):

$$\delta = \left(\frac{R_{SA} - R_{ST}}{R_{ST}} \cdot \delta_{SLAP} \middle/ \frac{R_{SLAP} - R_{VSMOW}}{R_{VSMOW}} \right), \tag{2}$$

with the additional term in the bracket being the normalization of the respective scale in terms of pre-defined hydrogen and oxygen isotope ratios of the two standards (table 1).

The two scales defined in equation (1) for SMOW and in (2) for VSMOW coincide only if $R_{SMOW} \equiv R_{VSMOW}$ and if the adopted δ-values for SLAP in table 1 correspond to the true ones as defined by equation (1) for both hydrogen and oxygen. From the reported measurements of NBS1 and VSMOW a slight offset of the zero-point of the two scales could be concluded (offsets of 0.05‰ and 0.5‰ for $\delta^{18}O$ and δ^2H, respectively), but which was well within the limits of measurement uncertainty of most laboratories.[19] Both offsets were a bit larger than evaluated before by Craig. However, due to the scatter of the individual results the stated mean offset should not be applied for conversion from one scale to the other.

A third water standard was proposed during the same meeting in 1976[19] with an isotopic composition intermediate between VSMOW and SLAP. This material was obtained from Greenland firn in 1978 and was called GISP (Greenland Ice Sheet Precipitation). GISP is intended to prove the successful calibration as performed with VSMOW and SLAP and the linearity of the measuring system. Results of two interlaboratory comparisons investigating GISP are published in IAEA reports.[20,7] It was noted that the accordance of results from different laboratories improved by a factor of more than two when the data were normalized using SLAP as second standard in addition to VSMOW.

2 INTENDED USE OF REFERENCE MATERIALS

2.1 Recent situation for IAEA RMs

The reference materials dedicated for the determination of stable isotope ratios at environmental level are distributed by the IAEA and on its behalf by NIST. A considerable number of units of these reference materials are distributed per year by the IAEA (figure 1) in spite of the established rigid rule on the distribution of these reference materials allowing only the order of one unit of each material per laboratory in a given three years period (see paragraph 2.1.5 below).

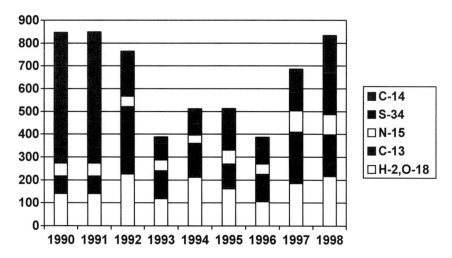

Figure 1 *Number of reference materials distributed by the IAEA Isotope Hydrology Unit
per year. These figures include all types of RMs and also those RMs which are
distributed free of charge (e.g.^{14}C), for calibration studies of RMs and for
interlaboratory comparisons.*

Most of these IAEA reference materials are described by Reports of Investigation
produced by the National Institute of Standards and Technology in USA. The IAEA
Isotope Hydrology Laboratory is in the process of updating the available information
according to the requirements as stated in ISO-Guide 31 and its recent revision.[21] In the
available NIST reports, the intended use of these materials is stated as in the following
example for VSMOW, SLAP, GISP: "These Reference Materials are intended to provide
samples of known isotopic composition with ^{2}H/^{1}H and ^{18}O/^{16}O isotope ratios stated in parts
per thousand difference (‰) from the VSMOW isotope-ratio standard. These RMs are not
certified, but their use allows comparability of stable hydrogen and oxygen isotope-ratio
data obtained by investigators in different laboratories. VSMOW and SLAP are intended for
stable hydrogen and oxygen isotope-ratio calibration of water samples and of mass
spectrometers for isotope-ratio analysis of all other oxygen- or hydrogen-bearing
substances."[22] A similar statement is provided for the other reference materials in other
issued NIST Reports of Investigation.

In the following sections, these IAEA reference materials will be discussed in respect to
the ISO Guide 33[23] recommendations.

2.1.1 Storage and Transfer of a property value.[23] Individual units of the water and gas
reference materials are stored in sealed 20 ml glass ampoules ready for distribution and
any remaining bulk material is stored in sealed ten liter glass containers. When opening a
sealed glass container to fill new samples for distribution, the isotopic composition of the
water reference material in the filled ampoules is always compared against ampoules filled
earlier from other containers. This cross-check verifies the proper storage of the materials
and excludes the possibility of any significant changes of the isotopic composition during
storage.

In case of the other solid reference materials they consist of materials which can be stored under ambient laboratory conditions in air-tight cap-sealed bottles and containers. In some instances the head space air is replaced by dry nitrogen or argon. Two carbonate materials are stored in desiccators under vacuum to prevent any reaction and isotopic exchange between the carbonate material, air moisture and air carbon dioxide.

2.1.2   Establishing traceability of the measurement result.[23] In case of the primary reference material VSMOW or the calibration materials NBS19 and IAEA-S-1 the traceability to SI units, in this case the mole, is achieved through the determination of the absolute isotope ratios of these materials using a definitive analysis method (isotope dilution method). In spite of this, until now these materials were not considered as certified RMs. One reason is the lack of information on some details of the formerly performed measurements (up to 28 years ago) required nowadays for a certification. In any case an absolute isotope ratio determination is quite useful, being an additional material-independent information which can facilitate in future the production of successor materials. However, the main purpose of these calibration materials is to give the means to the laboratory to trace their measurements back to an international accepted uniform δ-scale.

Virtually all other reference materials were established through interlaboratory comparison exercises. From a practical point of view, these RMs are in most cases the only available reference material for a specific type of substances and are therefore needed to calibrate any sample measurements versus a common scale. Some of these RMs should be considered as being traceable back only to the originating interlaboratory comparison exercise, especially those determined in the Seventies with a lack of available information on the formerly used methods, the laboratory calibration procedure and the uncertainty. This limitation should certainly not apply to these RMs which were tested only by few selected laboratories. In this case the traceability chain could be followed back directly to the primary reference material provided that these laboratories have given a detailed uncertainty budget for their measurements and sufficient information on the analysis procedures and provided all laboratory mean values and their uncertainties were taken properly into account for the final evaluation without a pure statistically driven outlier rejection. During the ongoing evaluation of the available reference materials their traceability will be re-assessed.

2.1.3   Determining the uncertainty of the measurement results.[23,24] The reporting of uncertainty of measurement results is requested from participants in IAEA interlaboratory comparisons. However, in many cases only some of the sources of uncertainty are stated like the standard deviation of individual determinations or no uncertainty statement is provided at all. Only a few laboratories comply by stating a total uncertainty budget for their measurement process. For future assessments of the suitability of reference materials only such laboratories should be considered which provide sufficient data on their measurement uncertainty. This would certainly eliminate some of the major constraints hindering the full use of interlaboratory comparison exercises for the characterization of new reference materials in terms of traceability. As long as the principle of an uncertainty budget is not implemented in all laboratories, a provisional approximation of the combined standard uncertainty for measurements could be the assessment of internal standards and their long term behaviour. The standard deviation calculated from such a time series should be quite useful for estimation of a laboratory uncertainty as expanded standard uncertainty.

2.1.4   Calibration of an apparatus and assessment of a measurement method. As pointed out, one of the main applications of the existing reference materials for stable isotope ratio assay is the calibration of equipment and measurement procedures. Due to the large spectrum of investigated substances in isotope hydrology and isotope geochemistry a variety of reference materials was produced to enable a calibration with a material as similar as possible.

2.1.5 Use of reference materials for quality control purposes. The available reference materials are intended to calibrate internal standards prepared by the individual laboratories. The RMs are NOT intended to be used themselves for quality control purposes. For the distribution of all stable isotope ratio reference materials a rather strict rule applies: Each laboratory is entitled to order one unit of any reference material only once in a three years period. This limitation was set to preserve the availability of the valuable reference materials for the maximal possible time and therefore to ensure the comparability of results from laboratories over long times.

2.2 Future Plans

Holding a key role in producing and distributing reference materials for the determination of ratios of stable isotopes in light elements, the International Atomic Energy Agency follows as much as possible the latest requirements for the production of reference materials and developments in the metrological part of isotope geochemistry. In regular intervals of two to three years Advisory Group and Consultants Meeting are convened to discuss the latest approaches and new developments in stable isotope reference materials and to focus the IAEA activities on the main priorities.

Recently developed analytical techniques enable laboratories to analyze sample amounts ten to hundred times smaller than some years ago on their isotopic composition (for example by gas chromatography isotope ratio mass spectrometry GC-IRMS, or the use of Elementar Analyzers for on-line combustion of samples). At the same time these methods challenge the established system of reference materials in view of the required homogeneity of the existing RMs for such small amounts, which was not taken in consideration at the time of their production. The Isotope Hydrology Unit therefore has started the re-evaluation of the existing materials in order to derive the best possible information on the former production steps and on the homogeneity and uncertainty of the established recommended values. This implies significant efforts for this Unit in order to carefully assess the established materials. This task should not be delegated to other laboratories in view of the problems of the lack of traceability and rigorous uncertainty assessment in many laboratories and in recognition of the limitations for experienced laboratories to dedicate themselves to time consuming and expensive tasks at an on principle cost-free basis. As part of the development of an quality assurance system for the IAEA laboratories the Isotope Hydrology Unit aims to update the existing protocols to comply to the requirements of the quality assurance system as well as to the necessary preconditions for the characterization and certification of reference materials.

2.2.1   Existing RMs for the assessment of stable isotope ratios in water samples. The existing know-how in the Isotope Hydrology Unit for high precision analysis of stable isotope ratios in water samples has already resulted in a project to produce a successor material for VSMOW. VSMOW is expected to be exhausted within the next seven to ten years. Since all measurements of hydrogen and oxygen isotope ratios base in principle on this material, the production of a suitable replacement has high priority. It was decided in accordance with the recommendation of an IAEA consultants meeting to initiate the production of a water indistinguishable in its hydrogen and oxygen isotopic composition from the existing VSMOW. In order to be able to also reproduce the ratios of ^{17}O and ^{18}O, isotopically enriched water could not be used to adjust its isotopic composition. Therefore three natural raw water were identified with their isotopic composition as close as possible to VSMOW but with slight differences therein to allow a triangular mixing approach to produce the final water (figure 2).

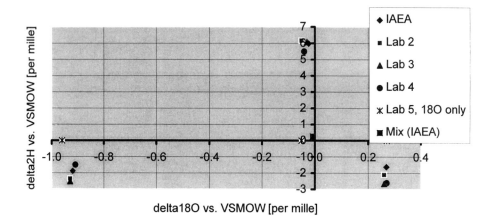

delta18O vs. VSMOW [per mille]

Figure 2   *Calibration of three raw waters for mixing a replacement for VSMOW by five laboratories (VSMOW being at the origin of the coordinate cross) and the preliminary result of a first trial mixing of the three components (big square close to the coordinate origin of the plot).*

The three raw waters were sampled, transferred to the Isotope Hydrology Unit and calibrated. This calibration was verified with the results of four other high precision stable isotope laboratories which were in accordance by better than ± 0.02‰ for $\delta^{18}O$ and by better than ±0.3‰ for $\delta^{2}H$ (see figure 2). In the next step the three raw waters will be mixed gravimetrically to produce the new standard which will then be tested carefully for any isotopic deviation from VSMOW.

Meanwhile the first attempts are initiated to obtain ice from the inner parts of Antarctica to produce a replacement for SLAP within the next years.

2.2.2   Existing RMs for carbon, sulfur and nitrogen isotopes. An intercalibration for the available sulfur stable isotope reference materials was started in 1998 in order to produce a consistent set of recommended values of all these RMs based on the VCDT scale. The evaluation of the results is ongoing.

A similar re-assessment is planned for some inorganic carbon RMs for 1999 in cooperation with NIST and invited laboratories. The Isotope Hydrology Unit has already upgraded its equipment to be able to test all existing inorganic carbon RMs and to produce new ones.

No similar activity is planned so far for nitrogen isotope RMs. The situation there seems to be satisfactory.

In view of the increasing importance of the determination of stable isotope ratios in organic materials (food authentication, medicine, biology, hydrogeology and isotope geochemistry), steps are foreseen to implement an appropriate analytical capability in the Isotope Hydrology Unit within the next two years for a high-precision determination of carbon isotopes in organic matter, mainly to be used for the certification of new reference materials.

  2.2.3   New reference materials to be produced in future. The analytical capabilities of the Isotope Hydrology Unit should be upgraded to enable it to determine the isotopic composition of all existing reference materials. So far, nitrogen and sulfur isotopes cannot be measured at all and the capability for carbon analysis in organic matter is just being developed. Due to the increasing constraints for other laboratories to perform measurements on IAEA reference materials for homogeneity tests on a cost-free basis and in order to keep control over the data evaluation, this effort seems to be necessary. The establishment of recommended values for future reference materials will not rely anymore on interlaboratory comparison exercises, but will be based on extensive measurements in the IAEA Isotope Hydrology Unit and the comparison of those results with a few selected laboratories of proven competence.

  2.2.4   Interlaboratory comparison exercises and proficiency tests. The main focus of such exercises in the future will be to support quality assurance in stable isotope laboratories. Proficiency tests will be performed with selected laboratories where appropriate and desired by using well calibrated internal standards stored and used in the Isotope Hydrology Laboratory.

  2.2.5   Distribution of reference materials. In general it would be desirable to change the distribution policy for future reference materials and allow laboratories to order these RMs more frequently. This implies that much larger batches of future RMs have to be prepared and handled. In the case of the successor material for VSMOW, about 200 litre will be produced, which is about three times more than previously for VSMOW.

References

[1] IAEA-TECDOC-825, Reference and intercomparison materials for stable isotopes of light elements, Proceedings of a Consultants Meeting held in Vienna, 1-3 Dec 1993, Sep 1995.

[2] A. Fajgelj, Z. Radecki, K. Burns, J. Moreno Bermudez, P. De Regge, J. La Rossa, P.R. Danesi, this issue.

[3] International Atomic Energy Agency, IAEA AQCS Catalogue for Reference Materials and Intercomparison Exercises 1998/1999, Vienna, 1998.

[4] K. Rozanski, Report on the Consultants' Group Meeting on 'C-14 Reference Materials for Radiocarbon Laboratories', held on 18-20 Feb 1991, Vienna, 1991.

[5] K. Rozanski, W. Stichler, R. Gonfiantini, E.M. Scott, R.P. Beukens, B. Kromer, J. van der Plicht, The IAEA 14C Intercomparison Exercise 1990, Radiocarbon, 1992, **34**, No.3, 506.

[6] 'International Vocabulary of Basic and General Terms in Metrology', International Organization for Standardization, 2nd ed., Geneva, Switzerland, 1993.

[7] L. Araguás Araguás, K. Rozanski, Report on 'Interlaboratory Comparison for Deuterium and Oxygen-18 Analysis of Precipitation Samples', International Atomic Energy Agency, Vienna, 1995.

[8] M. Magaritz and J.R. Gat, in: 'Stable Isotope Hydrology, deuterium and oxygen-18 in the water cycle' (J.R. Gat and R. Gonfiantini, Ed.), Technical Report Series No. 210, International Atomic Energy Agency, Vienna, 1981, Chapter 5, p. 85.

[9] R. Gonfiantini, in: 'Stable Isotope Hydrology, deuterium and oxygen-18 in the water cycle' (J.R. Gat and R. Gonfiantini, Ed.), Technical Report Series No. 210, International Atomic Energy Agency, Vienna, 1981, Chapter 4, 35.

[10] S. Epstein and T. Mayeda, Geochim. et Cosmochim Acta 1953, **4**, 213.

[11] H. Craig, Science, 1961, **133**, No. 3467, 1833.

[12] F.L. Mohler, Natl. Bur. Standards (U.S.), Tech. Note No. 51, 1960, 8.

[13] P. Baertschi, Earth and Planetary Science Letters, 1976, **31**, 341.

[14] W. Li, B. Ni, D. Jin, and T.L. Chang, Kexue Tongbao, Chinese Science Bulletin, 1988, **33**, 1610.

[15] R. Hagemann, G. Nief, and E. Roth, Tellus, 1970, **22**, 712.

[16] J.C. De Wit, C.M. van der Straaten, W.G. Mook, Geostandards Newsletter, **4**, no. 1, 33.

[17] R.S. Tse, S.C. Wong, C.P. Yuen, Analytical Chemistry, 1980, **52**, 2445.

[18] R. Gonfiantini, Standards for Stable Isotope Measurements in Natural Compounds, Nature 1978, **271**, No.5645, 534.

[19] R. Gonfiantini, Report on the IAEA Consultants' Meeting on 'Stable Isotope Standards and Intercalibration in Hydrology and in Geochemistry' at 8-10 Sep 1976, IAEA, Vienna, March 1977.

[20] R. Gonfiantini, Report on the IAEA Consultants' Meeting on 'Stable Isotope Standards and Intercalibration in Hydrology and in Geochemistry' at 8-10 Sep 1976, IAEA, Vienna, March 1977.

[21] ISO/REMCO No.488, 'Reference Materials - Contents of Certificates and Labels', Apr.1998, Revision of ISO Guide 31:1981, International Organisation for Standardisation.

[22] Report of Investigation, 'Reference Materials 8535-8537 VSMOW-GISP-SLAP', National Institute of Standards & Technology, Gaithersburg, USA, 15 Oct 1992.

[23] ISO Guide 33, 'Uses of Certified Reference Materials', Revision of ISO Guide 33: 1989, International Organisation for Standardisation, Geneva, Switzerland, 1998, to be published.

[24] 'Quantifying Uncertainty in Analytical Measurements', EURACHEM, First edition, 1995.

Case Study: Certification of Biological Reference Materials in the Czech Republic

Miloslav Suchánek

DEPARTMENT OF ANALYTICAL CHEMISTRY, INSTITUTE OF CHEMICAL TECHNOLOGY,
TECHNICKÁ 5, CZ-166 28 PRAGUE 6, CZECH REPUBLIC

Pavel Mader

DEPARTMENT OF CHEMISTRY, CZECH UNIVERSITY OF AGRICULTURE, CZ-165 21
PRAGUE 6, CZECH REPUBLIC

1 INTRODUCTION

While the history of preparation and certification of geological and metallurgical reference materials in former Czechoslovakia goes back to the sixties, materials with biological matrix have been prepared only recently. The first of them, the „CRM 12-02-01 „*Bovine Liver*", was prepared at the Czech University of Agriculture in Prague in 1989.[1] It was soon followed by four materials of plant origin, namely CRM 12-02-02 „*Green Algae*", CRM 12-02-03 „*Lucerne*", 12-02-04 „*Wheat Bread Flower*", and CRM 12-02-05 „*Rye Bread Flower*".[2-5] In all cases, essential and risk elements were the analytes followed. Basic information about these materials has been given also at the BERM-5 symposium.[6,7] Two more candidate RMs with animal matrix (*Bovine Kidney, Bovine Muscle*) have also been prepared, underwent an interlaboratory comparison and certified values have been proposed.[8] However, in connection with the division of Czechoslovakia and the fact that majority of the former Federal Institute of Metrology remained on Slovak territory, their certification campaign has not been completed.

The newly established Czech Metrology Institute has ascribed the CRMs its new codes. Till today, two new CRMs with the biological matrix have been certified, namely CRM 6001 „*Creatine in Human Urine*", and CRM 6002 „*Horse Liver - Risk Elements*". We have chosen this latest CRM for the case study which is subject of this contribution.

2 CASE STUDY I

2.1 Preparation, certification, and use of reference material: hazardous elements in horse liver

Reference material mentioned seems to be useful for the analysis of As, Cd, Cu, Hg, Pb and Zn in biological matrices with high content of risk elements. The interesting application is also supposed for the analysis of human organs. Preferred use to be recommended is for validation of analytical procedures for the determination of several hazardous elements, and for calibration of such validated methods.

Preparation and certification of reference material (RM) consisted of following steps:
1. collecting livers from horses in a northern district of the Czech Republic
2. storing horse livers samples at temperature below -27°C
3. homogenization of liver samples by cutting into small pieces
4. freeze drying
5. grinding of freeze dried liver to liver dust
6. sieving of liver dust
7. homogenization in inert atmosphere (nitrogen)
8. filling of homogeneous material to PE bottles, labelling, sealing to aluminium foil and again labelling
9. homogeneity testing
10. stabilization of material by γ-radiation
11. stability testing
12. certification campaign

2.2 Detailed preparation procedure

The CRM candidate material was collected in the northen Moravia district slaughterhouse over a period of one year. Each tissue was frozen immediately after collecting and stored below -27°C, till the overall target amount (aprox. 100 kg) was achieved. The whole bulk of candidate material was then transported refrigerated for freeze drying, grinding, blending and sieving to obtain fraction below 300 μm. The resulting material was adjusted to the PE bottle units and sealed and radiation sterilised by the total dose of 25 kGy of γ-radiation. Finally the subsamples were taken for the homogeneity testing.

2.3 Homogeneity testing

Both within-unit and between-unit homogeneity was tested according to the ISO Guide 35 for all certified elements by methods with a sufficient repeatability. Homogeneity was then compared to expected uncertainty of the certified values. The inhomogeneity was found to be insignificant for all elements at the confidence level $\alpha = 0.05$, for the minimum analytical subsample mass of 0.5 g.

2.3 Stability testing

The stability of RM was tested and found to be satisfactory. The stability is guaranteed for the whole validity period (until 2002) provided the storage requirements are fulfilled (see below)

2.4 Certification campaign

15 Czech laboratories participated in the interlaboratory certification experiment. The following analytical techniques were applied: FAAS, ETAAS, HGAAS, DPASV, ICP OES. The calibration of methods used was made with the use of standard solutions of metals mentioned. Standardization of standard solutions was done by primary methods of measurement (titrimetry, mainly by complexometric titration). The expanded uncertainty is

expressed as the 95% confidence interval based on the variation of laboratory means. Certified and noncertified values are shown in Table 1.

Table 1 *Certified and Non-certified Values of CRM Horse Lliver*

Element	Certified value/mg kg^{-1}	Expanded uncertainty/mg kg^{-1}
As	0.0298	0.0035
Cd	32.94	1.80
Cu	91.0	4.5
Hg	0.027	0.001
Pb	0.804	0.083
Zn	228.1	5.4
Al	8.00*	
Cr	0.15*	
Ni	0.24*	

* non-certified values

2.5 Instructions for proper use

The RM should be kept tightly closed in its original bottle and stored in the dark at a temperature below 20°C. The bottle with RM should be shaken properly prior to every use of the sample for determination to eliminate any possible segregation. The certified values are expressed for the dry material. The user should perform the dry matter determination parallel to the element determination. This may be done by drying 200 mg of RM sample at 102°C till constant mass, i.e. the difference between two consecutive weighing below 0.2 mg, and to calculate the values found for element to the dry matter so determined. The dried material must not be used for the element determination.

2.5.1 Recommended procedure for decomposition. Prior to the instrumental measurement the RM should be decomposed. Recommended procedures for decomposition are given below.

2.5.2 Dry decomposition: Furnace, small amount of nitric acid as a matrix modificator, controlled increasing temperature from 90 °C to max. 470 °C, duration 12 hrs. Resulting ash can be dissolved in 10 ml of 1 mol/l nitric acid.

2.5.3 Wet decomposition: Microwave decomposition, nitric acid and hydrogen peroxide.

2.6 Verification of calibration method

Calibration of instrumental method can be done with the use of standard solution of respective metal. Concentration and/or content of metal should be checked by primary method of measurement, e.g. gravimetry (Cu, Al, Ni), titrimetry (Cd, Hg, Zn, Ni, Cu) or coulometry (As, Cr). Accuracy of calibration used is suggested to be verified by analysis of RM Horse Liver. Accuracy of the result cover also recovery of matrix decomposition. The results of verification for various metals in one laboratory are shown in Table 2.

With the exception of Pb, the laboratory reached satisfactory results for verification of instrument calibration.

Table 2 *Verification of Calibration of Instrumental Method*

Element	Method	Certified value/mg kg^{-1}	Verification results/mg kg^{-1}
Hg	cold vapour	0.027±0.001	0.0284±0.0009[*]
Cu	FAAS	91.0±4.5	87.90±3.27[*]
As	HGAAS	0.0298±0.0035	0.0277±0.0040[*]
Cd	FAAS	32.94±1.80	34.70±0.55[*]
Pb	FAAS	0.804±0.083	0.668±0.031[*]
Zn	FAAS	228.1±5.4	224.3±2.3[*]

* 95% confidence interval (from 5 independent repetition of measurement)

3 CASE STUDY II: DETERMINATION OF SOME HEAVY METALS IN RYE GRASS BY ICP/MS

Accuracy, limit of detection, and uncertainty budget of standard operational procedure for determination of some heavy metals were evaluated. As biotic matrix, CRM 281 Rye Grass (BCR No. 561) was used. CRM was decomposed by microwave technique, resulting solution was measured by ICP/MS method. Below in Tables 3 and 4, some technical parameters of microwave furnace and parameters for decomposition are presented.

Table 3 *Technical Parameters of Microwave Furnace BM-1S/II, Plazmatronika (Poland)*

power supply	220V ± 20V; 50 Hz
energy input	600 W
microwave energy per one vessel	150 W
total volume of vessel	110 cm^3
cooling by water - min. pressure	0.5 Mpa
ventilation - air flow rate	3 m^3. min^{-1}
total weight	33 kg

Table 4 *Decomposition Program*

sample:	500 mg
reagent:	3 ml 65% (m/m) HNO$_3$
total decomposition time:	10 min
cooling time	10 min
pressure limit	25 atm
max. output	300 W

Calibration of ICP method was done with the standard solutions of respective metal. Concentration of standard solutions and its uncertainty (expressed as combined uncertainty) was determined by primary method of measurement.

As an example, determination of Ni concentration in standard solution and evaluation of uncertainty of concentration is presented as follows. Nickel was determined by primary method of measurement - gravimetry with nickel-2,3-butandiondioximate as resulting precipitate. Concentration of standard solution of Ni is calculated according to the formula

$$c_{Ni} = \frac{m_v \cdot M_{Ni} \cdot 10^3}{M_{prec} \cdot V}$$

where m_v is weight of precipitate (nickel-2,3-butandiondioximate), M_{Ni} and M_{prec} are molar weights of Ni and precipitate, V is pipetted volume of standard solution (25 ml). Standard uncertainties of individual quantity are shown in Table 5:

Table 5 *Standard Uncertainties of Some Quantities*

Quantity	Value	Standard uncertainty	RSD
m_v	129.5 mg	0.281 mg	0.00217
V	25 ml	0.019 ml	0.00076
M_{Ni}	58.69 g mol^{-1}	0.00012 g mol^{-1}	0.000002
M_{prec}	288.91 g mol^{-1}	0.0038 g mol^{-1}	0.000013

Standard uncertainty of precipitation was estimated from conditional solubility product of Ni-biacetyldioxime the value of which is approximately 10^{-16}. Concentration of reagent (biacetyldioxime) in solution is 0.003 mol l^{-1}, mass of Ni remaining in solution after precipitate separation is 1.3×10^{-10} g, estimated relative standard deviation is approx. 5×10^{-9}. This value was omitted in next uncertainty evaluation.

Combined standard uncertainty of Ni determination is 0.0026 g l^{-1} (RSD 0.0025), resulting concentration of Ni standard solution is:

$$c_{Ni} = (1.0523 \pm 0.0052) \text{ g l}^{-1}$$

Repeatability standard deviation was estimated from nine repetitive determinations. RSD was found to be 0.00085 and is more than two times lower than combined standard uncertainty.

3.1 Calibration plan

In Table 6, concentration of calibration solutions over the linear range is done. Solutions S0 and S3 were measured 10 times, measurement with solutions S1 and S2 was not repeated. To all solutions, Bi and In were added as internal standards. All solutions were prepared by dilution of standard solutions.

Table 6 *Calibration Plan*

Element	Calibration solution/$\mu g\ ml^{-1}$			
	S0	S1	S2	S3
Zn	0	100	200	500
Cr,Cu,Ni,Pb	0	20	50	100
Cd	0	2	5	10
Bi,In	100	100	100	100

3.2 Uncertainty budget

Metal content in CRM is calculated according to the formula:

$$w_M = \frac{(c_{sample} - c_{bl}).V.f}{m(100 - w_{H_2O}).10}$$

where w_M is content of metal in CRM, c_{sample} and c_{bl} are concentrations of metal in analyzed solution and blank, resp., calculated from calibration curve, m is sample weight, f is dilution factor, V is sample volume (50 ml), and w_{H_2O} is water content in CRM. Uncertainty budget consists of uncertainty of decomposition and uncertainty of calibration. Standard uncertainties of some quantities are in the Table 7.

Table 7 *Standard Uncertainty of Some Quantities*

Quantity	Value	Standard uncertainty	RSD
m	0.5060	0.073 mg	0.00014
V	50	0.061 ml	0.0012
f	2	0.0019	0.00096
w_{H_2O}	7.31%	0.11%	0.015

Standard uncertainty (as RSD) of decomposition was varied from 0.0023 to 0.026 for all metals analyzed. Standard uncertainty of calibration was varied from 0.0024 to 0.090 also for all metals. In Table 8 content of metal isotopes with expanded uncertainty are shown.

Table 8 *Result of Analysis of CRM Rye Grass*

Isotop	Result	Certified value
^{53}Cr	2.20±0.24	2.14±0.12
^{62}Ni	3.36±0.48	3.00±0.17
^{65}Cu	10.33±0.43	9.65±0.38
^{70}Zn	33.1±1.3	31.5±1.4
^{111}Cd	0.125±0.024	0.120±0.003
^{208}Pb	2.48±0.30	2.38±0.11

References

1. J. Kučera, P. Mader, D. Miholová, J. Cibulka, M. Poláková, D. Kordík: Report on Interlaboratory Comparison of the Determination of the Contents of Trace Elements in Bovine Liver 12-02-01. Czechoslovak Institute of Metrology, Bratislava, 1989.
2. Š. Bartha, M. Kalinčák, D. Kladeková: Report on Intercomparison for the Determination of Essential and Toxic Elements in Green Algae. Institute of Radioecology and Applied Nuclear Techniques, Košice, 1989.
3. Š. Bartha, M. Kalinčák, D. Kladeková: Report on Intercomparison for the Determination of Essential and Toxic Elements in Lucerne. Institute of Radioecology and Applied Nuclear Techniques, Košice, 1989.
4. M. Kalinčák, Š. Bartha, D. Kladeková: Report on Intercomparison for the Determination of Essential and Toxic Elements in Whear Bread Flower. Institute of Radioecology and Applied Nuclear Techniques, Košice, 1991.
5. M. Kalinčák, Š. Bartha, D. Kladeková: Report on Intercomparison for the Determination of Essential and Toxic Elements in Rye Bread Flower. Institute of Radioecology and Applied Nuclear Techniques, Košice, 1991.
6. D. Miholová, P. Mader, J. Száková, A. Slámová, Z. Svatoš: Czechoslovakian Biological Certified Reference Materials and their use in the Analytical Quality Assurance System in our Trace Element Laboratory. - In: Book of Abstracts. 5th International Symposium on Biological and Environmental Reference Materials. Aachen, May 11-14, 1992, p. 83.
7. D. Miholová, P. Mader, J. Száková, A. Slámová, Z. Svatoš, *Fresenius J. Anal. Chem.* 1993, **345**, 256.
8. J. Kučera, P. Mader, D. Miholová, J. Cibulka, J. Faltejsek, D. Kordík, *Fresenius J. Anal. Chem.* 1995, **352**, 66.

Certification of Method-Defined Parameters in Matrix-CRMS: The Preparation and Use of Coal Reference Materials

Adriaan M. H. van der Veen

NEDERLANDS MEETINSTITUUT, DEPARTMENT OF CHEMISTRY, PO BOX 654, 2600 AR DELFT, THE NETHERLANDS

1 INTRODUCTION

In its position as National Metrological Institute, NMi Van Swinden Laboratorium B.V. has a long-term experience in the production and certification of measurement standards for physical and chemical measurement. Within the Department of Chemistry, Primary Standard Mixtures (PSMs) are established that are traceable to SI. These PSMs are synthetic gas mixtures, prepared by means of gravimetry. The composition of these mixtures is calculated from the preparation data and the purity of the parent gases, and then verified. Apart from the PSMs, the Department of Chemistry also produces diffusion tubes with (for instance) volatile organic compounds for ambient atmosphere measurements.

In the former Department for Solid State Reference Materials[1], particulate materials were prepared as basis for the production and certification of solid state reference materials. A long-term experience existed in the process of grinding and subdividing of particulate materials in order to prepare standard samples. With the project "Preparation and characterisation of coal samples and maceral concentrates for studies on gasification and combustion reactivity of coals in combined cycle processes" (ECSC 7220/EC-036) [1] the possibility of preparing certified quality control reference materials was investigated. A series of certified reference materials for coal analysis resulted.

The scheme of traceable gas mixtures was taken as the starting point for setting up the programme. As the materials were to be certified on the basis of interlaboratory studies, special precautions had to be taken in order to ensure comparability between the results obtained on different batches of samples from different coals. This paper mainly deals with the relationship between the details of the production and certification of the materials, and the consequences for use. There is a clear connection between the certification procedure chosen by the producer and the possible applications of the CRM (certified reference material). One of the key issues is the uncertainty assigned to the certified values. This uncertainty statement is based on several assumptions and is based on either (1) the certified value or (2) any measurement carried out on the material, or any group of measurements.

---

[1] This department is now a stand-alone company, "Referentiematerialen Nederland B.V.".

2 BACKGROUND

Coal is still one of the most important energy vectors in the world. The material is used on a large scale in the generation of electric power and in the production of steel. Besides these two main applications, there are a wide variety of other applications where coal is used. Coal is traded on a world-wide basis. The economic value of coal is determined based on coal analyses. When seller and buyer reach an agreement on the properties of a specific lot, the transaction can take place. This agreement is based on results from two laboratories, one representing the buyer, the other representing the seller.

In coal trade, the comparability of measurement results is of primary importance. In those cases where no agreement can be reached based on the results of the two laboratories, the transaction is delayed and extra analyses are required. Sometimes even the involvement of a third laboratory is required, based on the contract. For an efficient transaction however, it is crucial for laboratories to be able to work under well defined, traceable conditions. Usually this traceability of measurement results is established by using a certified reference material that is representative for the range of coals usually analysed.

In a project funded by the European Coal and Steel Community (ECSC), a suite of coals were prepared and characterised both for establishing comparability between laboratories as well as providing materials that could be used for method verification and research. For coal research, often specially prepared samples are needed. One of the aims of this paper is to demonstrate how this can be done in practice, and what practical implications exist.

The main objective of this paper is to show how, within the framework of ISO Guide 35 [2], matrix certified reference materials can be prepared and certified. From the above mentioned project, one coal has been selected that may be regarded to be representative for all coals in the project.

3 SETUP OF THE INTER-LABORATORY STUDY PROGRAMME

Over a period of three years, an interlaboratory study programme was organised that aimed to supply the coal community with a range of suitable reference materials, that can be used as measurement standards in a wide variety of experiments and common analyses. The programme was entitled "ILS Coal Characterisation" and consisted of 8 interlaboratory studies. Apart from the objective of supplying measurement standards, it was also aimed to investigate several metrological aspects, such as establishing traceability of measurement results throughout the programme. Summarising, the interlaboratory study programme aimed to [1]

1. establish a series of well characterised coals samples (reference materials) in order to support coal research
2. investigate the statistic parameters of coal analysis
3. study the influence factors such as sample preparation, sub sampling, and statistics on measurement results

The interlaboratory studies have been evaluated and reported during the programme [3-10]. After completion of the programme, a report was published that covers the evaluation of the interlaboratory study programme as a whole [1]. This second evaluation cycle aimed to establish links between interlaboratory studies and thus establishing performance characteristics for selected methods common in coal analyses throughout the programme. These performance characteristics are believed to be better established than those from a single interlaboratory study, obtained with the procedure given in ISO 5725:1994, parts 1 and 2 [11,12].

4 SAMPLE SELECTION AND SAMPLING

Most materials for the programme were sampled as "run of mine" coals (ROM-coals). A ROM-coal is a coal as a mine produces it. It may be a single seam coal, but more often it is a blend of several seams in order to ensure a certain level of quality. The Rietspruit coal, which will be taken as an example in this paper, was sampled after transport from South Africa to the Netherlands. A batch of about 1 000 kg was obtained, in order to allow to produce in addition to the certified reference material a sufficient number of samples of different grain sizes and different amounts for research. The latter samples were not certified, but the production process had to be developed in such a way, that the values of the CRM were valid for the research materials as well.

Often coals of different mines are blended, and in the programme "ILS Coal Characterisation" several of these blends were used as well. From the Göttelborn coal [3,5] for instance, two batches of materials were prepared and used in different rounds of the programme. From the results obtained on these coals, it could be derived that the preparation procedure as used for both the candidate-CRMs and for the research samples allowed to use the values from the certification interlaboratory study for the research samples as well.

5 PREPARATION OF THE MATERIALS

The preparation of the materials started with drying them at room temperature, or at a temperature not higher than 40°C. Higher temperatures would require to dry these amounts of coal (typically about 1 000 kg) in nitrogen in order to prevent oxidation. Apart from the problem of constructing a room that can be evacuated and filled with nitrogen, there is a second problem with drying coal at elevated temperatures, in fact loss of volatile matter. Depending on the rank of the coal, the content of volatile matter is higher or lower. Even at a temperature of 40°C, with low–rank coals one can smell the volatile matter of the coal. As these are usually aromatic species, it is easily observed whether there is a serious loss of volatile matter.

Subsequent to drying, the particle size had to be reduced. For traded coals, the top size is already reduced to 50 mm particle size. For reference materials, the top size is to be reduced to 0.200 mm. This can only be done in a step-wise process, as crushing and milling have a potential side-effects the production of heat, with similar consequences for the material as drying at elevated temperatures. In this case, the consequences are even worse, as the freshly formed surface of the particles is usually very reactive, whereas the surface of a sampled coal is usually deactivated due to ageing.

In order to reduce the side-effects of milling, milling took place under cryogenic conditions. Liquid nitrogen was used a cooling agent. Special care had to be taken when cooling the coal, as there is the bad wetting of coal with liquid nitrogen and there is the violent vaporisation of it during the cooling process. With an optimised procedure of adding liquid nitrogen, in combination with sieving off the fractions already sufficiency ground, the loss of material could be reduced to an absolute minimum.

The process of preparing research samples has been sketched in figure 1.

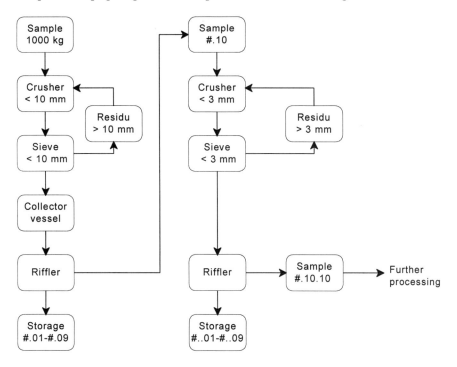

Figure 1 *Preparation scheme for standard samples [1]*

The riffler appearing in the box in the left-most.column is a specially designed device for handling batches between 100 and 1000 kg. The riffler is equipped with a rotating head, that subdivides the flow of material over 10 tubes. At the initial level of 1000 kg starting material as well as on the second level of 100 kg, representative samples under strict controlled conditions can be prepared. In this way, the preparation procedure allows to use the results achieved on a small subset of samples from the material for the whole batch, as long as the subsample is still representative for the batch. In principle, this representativeness cannot be guarantied any more after (1) manual subsampling and (2) fractionation by means of e.g. sieving.

The basic procedure for obtaining representative subsamples from a bulk sample is outlined in figure 1. The sample passes a jaw crusher, a sieve, the collector vessel, and the spinning riffler. The riffling procedure is carried out with a decimal system, for practical reasons. This implies that all riffling equipment has been modified accordingly.

The selection of the 10th sample is arbitrary: since every sub sample is supposed to have identical properties, any sample could have been chosen. However, selecting systematically the 10th sample each time influences the statistics. Since the likelihood of having another sample is 0, at least the bias of the riffler (which may or may not be significant) is transferred into this sample. This bias may be limited to the (relative) amount of material in subsample #.10, but it may also influence the constitution of the subsample. However, in practice the differences between the subsamples at this stage of subdividing are small enough. The difference between groups of samples was investigated in this interlaboratory study programme with the Göttelborn coal in ILS Coal Characterisation I and III, as already mentioned.

If a reference material is to be produced, one of the nine remaining samples is taken, and prepared as a reference material. An outline of this procedure is shown in figure 2. This procedure has been applied to selected coals in the project for the interlaboratory study programme. The principle of the procedure applied is the same as for the preparation of reference materials.

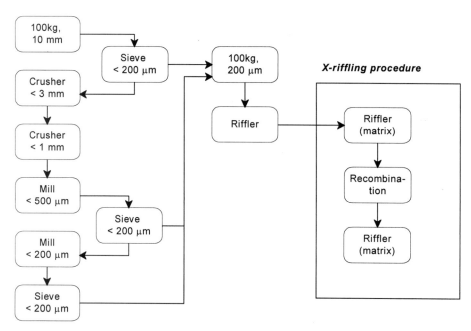

Figure 2 *Preparation scheme of a reference material [1]*

From figure 2, it can be seen that the reduction of the particle size (left-most column) is a process of several steps. Sieving is applied whenever necessary, but usually more than once in the process of grinding. The operator makes this judgement, so the use of a sieve may differ from one coal to another. The final sieving step is always required however. Although not shown, the material not passing the 200 μm sieve at the bottom of the first column is fed to the mill once more.

After the complete sample has been ground, all fractions are combined, and then riffled with use of a spinning riffler. This step produces 10 sub samples of about equal properties. Since this step is normally carried out on an equipment which has a lower accuracy than a laboratory riffler, the cross-riffling (X-riffling) procedure is applied. This procedure has proved to be effective for reducing the differences in mass after riffling [6] and cancelling out systematic effects from earlier steps in the production process. Usually the material was first well-homogenised by mixing before starting the X-riffling process.

The preparation of a large number of identical sub samples requires a sub sampling procedure which guarantees that each of the samples produced are as identical as possible. In the previous section, the use of a dynamic riffler has been recommended to do so. The principles of X-riffling do not require the use of a dynamic riffler, but they are only valid if the precision of the process used for X-riffling is better than that of the preceding step. The procedure developed during the years 1990-1995 for the preparation of large number of identical samples is called cross-riffling (X-riffling). It should be noted that X-riffling is just a method to split and recombine samples in order to cancel out effects of previous riffling steps.

The first step in the X-riffling process is to sub divide the material into 10 portions. After riffling of the starting material, each of the 10 portions (labelled #01 .. #10) is riffled into 10 sub samples. The next series of samples are being placed in such a way, that in every row, the first pair of digits as well as the second pair of digits are *unique*. The most convenient way to proceed is to have the second pair of digits in ascending (or descending) order in a row. After number #.10 it follows #.01. For the principle of X-riffling, this is not important, but for the operator, it is. After completion of the 10 riffling steps which build this matrix, it is easily checked whether each of the samples to be created by recombination of a row will contain one sample from each series (first pair of digits) and one sample from each tube of the riffler (second pair of digits).

The complete matrix after riffling the samples #.01 through #.10 reads as follows

01	02	03	04	05	06	07	08	09	10	
⇓	⇓	⇓	⇓	⇓	⇓	⇓	⇓	⇓	⇓	
01.01	02.02	03.03	04.04	05.05	06.06	07.07	08.08	09.09	10.10	⇒
01.02	02.03	03.04	04.05	05.06	06.07	07.08	08.09	09.10	10.01	⇒
01.03	02.04	03.05	04.06	05.07	06.08	07.09	08.10	09.01	10.02	⇒
01.04	02.05	03.06	04.07	05.08	06.09	07.10	08.01	09.02	10.03	⇒
01.05	02.06	03.07	04.08	05.09	06.10	07.01	08.02	09.03	10.04	⇒
01.06	02.07	03.08	04.09	05.10	06.01	07.02	08.03	09.04	10.05	⇒
01.07	02.08	03.09	04.10	05.01	06.02	07.03	08.04	09.05	10.06	⇒
01.08	02.09	03.10	04.01	05.02	06.03	07.04	08.05	09.06	10.07	⇒
01.09	02.10	03.01	04.02	05.03	06.04	07.05	08.06	09.07	10.08	⇒
01.10	02.01	03.02	04.03	05.04	06.05	07.06	08.07	09.08	10.09	⇒

The second pair of digits should also be unique, in order to prevent propagation of the riffler bias. Although laboratory rifflers are very accurate, it cannot be excluded that the riffler has a systematic error (selectivity). This selectivity may cause two unwanted effects: a systematic difference in amounts of material after riffling or a systematic change in particle size distribution. Now the samples of each row are put together, yielding 10 sub samples with identical properties, labelled A..J

01	02	03	04	05	06	07	08	09	10
⇓	⇓	⇓	⇓	⇓	⇓	⇓	⇓	⇓	⇓

01.01	02.02	03.03	04.04	05.05	06.06	07.07	08.08	09.09	10.10	⇒	A
01.02	02.03	03.04	04.05	05.06	06.07	07.08	08.09	09.10	10.01	⇒	B
01.03	02.04	03.05	04.06	05.07	06.08	07.09	08.10	09.01	10.02	⇒	C
01.04	02.05	03.06	04.07	05.08	06.09	07.10	08.01	09.02	10.03	⇒	D
01.05	02.06	03.07	04.08	05.09	06.10	07.01	08.02	09.03	10.04	⇒	E
01.06	02.07	03.08	04.09	05.10	06.01	07.02	08.03	09.04	10.05	⇒	F
01.07	02.08	03.09	04.10	05.01	06.02	07.03	08.04	09.05	10.06	⇒	G
01.08	02.09	03.10	04.01	05.02	06.03	07.04	08.05	09.06	10.07	⇒	H
01.09	02.10	03.01	04.02	05.03	06.04	07.05	08.06	09.07	10.08	⇒	I
01.10	02.01	03.02	04.03	05.04	06.05	07.06	08.07	09.08	10.09	⇒	J

If 100 samples are needed, the samples A..J are riffled again, yielding 100 sub samples with identical properties. For other quantities (1000 samples for instance), each of the samples A..J is riffled, followed by riffling each of the 100 sub samples. It could be argued that the process of X-riffling should be applied as final step in the procedure. From a statistical modelling carried out in 1993, it has been demonstrated that this is not necessary, as long as the subdividing process is under statistical control.

6 PARAMETERS TO BE CERTIFIED

The parameters to be certified comprise all parameters relevant to the economic value of coal: moisture, volatile matter content, ash content, gross and net calorific value, carbon, hydrogen, nitrogen, sulphur, chlorine and ash composition. The net calorific value mainly determines the economic value of a coal. It depends on the moisture content, gross calorific value, and hydrogen content. If it is to be calculated on an ash-free basis then the ash content is also involved.

For the characterisation of coal, there are two series of written standards, ASTM and ISO. In this interlaboratory study programme, the ISO-standards were requested. The laboratories were allowed to use their own methods if the results of these methods are comparable to those obtained with the ISO-method. It was the responsibility of the laboratory to verify whether their method is comparable to the ISO-method.

Several methods, such as the determination of the ash content, define the parameter: ash is the result of a chemical conversion of coal. Its formation (and as a result, the ash content) highly depends on the conditions under which the coal is burnt. This fact presents some consequences. The first consequence is that it is generally not possible to determine the parameter with independent methods. As a result, the best realisation of the parameter depends on how closely the written standard is followed by the participants and how many freedom is still left in the measurement method. Traceability of the parameter is limited to the measurement results being traceable to the written standard.

The evaluation protocol was merely based on a combination of ISO 5725-1:1994 [11], ISO 5725-2:1994 [12], ISO Guide 43[2] [13], and ISO Guide 35:1989 [2]. Outliers and/or stragglers [12] were identified by computing a Z-score [13], based on the mean and the standard deviation of the laboratory averages. The criterion was that this "Z" should not exceed a value of 2. The criterion was developed based on requirements set by all

---
[2] During the interlaboratory study programme, the most recent draft of this ISO Guide was used.

parties involved in coal production, trade, and consumption. For all samples involved in the programme, the performance characteristics were computed after removal of stragglers and outliers. A database of values of grand mean, repeatability standard deviation, and reproducibility standard deviation resulted.

7 HOMOGENEITY TESTING

The homogeneity of the batch of samples was not verified prior to the interlaboratory study. Checking of it took place by investigating the results of the laboratories. Each laboratory received three bottles, of which one value was to be reported. So, at the level of each laboratory, the standard deviation obtained addressed the between-bottle homogeneity. Prior to sending out the samples, some experiments were run with a thermogravimeter (TGA), in order to study the combustion profile. From these profiles, the ash content could be calculated, and by this, it could be verified that by repeating measurements that the samples were sufficiently homogeneous at a level of at least 40 mg, which was the maximum sample intake of the TGA.

As all methods for coal analyses require about 1 g minimum sample intake or more, there were never issues like insufficient homogeneity found. The preparation of these certified reference materials demonstrates clearly how long-term experience in handling a particular matrix can be utilised in addressing issues like homogeneity and stability testing. In many areas, a separate homogeneity test is required. Especially in trace analysis, an experimental set-up like this one is insufficient, as differences between samples from the same type of matrix may differ considerably. Coal is a relatively well described matrix in literature. As such, it is a material that allows to develop procedures for utilising context-knowledge in the preparation and certification of reference materials.

8 STABILITY TESTING

Stability testing was set-up by using these combustion profiles. From literature, it was known that changes in these profiles were the first indication of degradation of the matrix. It does –apart from the ash content- not provide any quantitative results on the certified parameters, but the kinetics of combustion are far more sensitive towards ageing than the certified parameters. These measurements were carried out on a regular basis, depending on coal rank in time intervals of 6 months up to 12 months. Anthracites (high-rank coals) were investigated less frequently than the middle-rank coals (like Rietspruit) and the low-rank coals.

In order to ensure stability, the materials are stored in dark glass bottles of 100 ml, filled with nitrogen and closed air-tight. The user is recommended to use the CRM a limited time after opening it for the first time.

9 IMPLEMENTATION OF TRACEABILITY TO OTHER COAL CRMs

In each interlaboratory study, a blind sample was used, except in round I [3]. The link between the results of this interlaboratory study and the results of the other rounds in the programme was established by using the materials of ILS Coal Characterisation I in later rounds [1]. Figure 1 shows the principle of establishing these traceability links in an interlaboratory study programme. The use of a blind sample enables to evaluate whether the results of a round are comparable to those of other interlaboratory studies due to the fact that each laboratory was requested to perform the measurement of all samples of the suite as independent measurements under repeatability conditions.   This way of implementing comparability enables also the assessment of traceability of measurement results to the written standard. All laboratories involved use internationally accepted certified reference materials as a part of their QAS. So, from that point a traceability link is established between these reference materials and the results of the interlaboratory study programme.

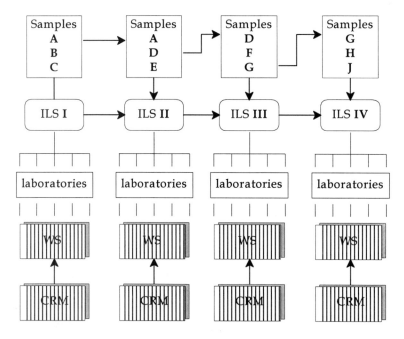

Figure 3 *Implementation of traceability in the interlaboratory study programme*

Based on the evaluation of the results of the blind samples, it may be concluded that generally speaking there is a good agreement between results obtained in two interlaboratory studies on a single batch of samples [4-10,1]. An example of it will be given in the next section.

These results lead to the possibility of establishing fairly homogeneous performance characteristics for the coal analyses involved in the programme [1]. It also lead to some inferences about the relationship between the repeatability standard deviation and reproducibility standard deviation as obtained from ISO 5725-2 and the concept of measurement uncertainty [14]. This analysis has been published elsewhere. [15]

10 ASSESSMENT OF TRACEABILITY

As an illustration of how the scheme of figure 3 was implemented, the results of two parameters are presented below. Rietspruit (267SA25) was used twice in the interlaboratory study programme, in fact in ILS Coal Characterisation VI and ILS Coal Characterisation VII. The coal served as a blind for the parameters of proximate and ultimate analysis in ILS VII. The parameters selected are gross calorific value, which is relatively insensitive for subsampling problems and ash content that is.

The second time a coal was used in the programme, it was used as a blind sample, and labelled accordingly. No information to the participants was submitted about the blind sample. So, the results of the blind sample may be regarded as being truly independent from the first characterisation round.

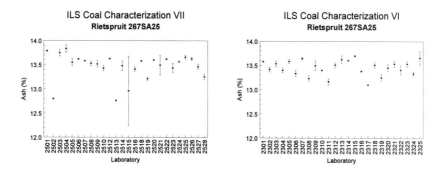

Figure 4 *Results for ash content (weight %) in ILS VII (left) and ILS VI (right) on Rietspruit*

The results on the ash content of Rietspruit are shown in figure 4. The standard deviation of laboratory 2515 (ILS VII) is too large. This value has been flagged as an outlier, as there is no indication from the other laboratories, that some inhomogeneity between sample bottles exist. Furthermore, laboratories 2502 and 2513 have been marked as outliers based on their Z-score. After rejecting the results of these three participants, the grand mean is 13.55, the repeatability is 0.15, and the reproducibility is 0.42. These values match very well with those obtained in ILS Coal Characterisation VI (13.48 (grand mean), 0.18 (repeatability) and 0.43 (reproducibility). In ILS VI one outlier was found: 2317.

The results for gross calorific value are shown in figure 5. Laboratory 2502 is marked as an outlier (ILS VII). The grand mean is 28.47, the repeatability is 0.12, and the reproducibility is 0.33 (MJ/kg). These performance characteristics agree very well with the limits in the ISO 1928 standard. In ILS VI the results of laboratory 2311 have been rejected as outlier. After correction, the grand mean is 28.51, the repeatability is 0.12 and the reproducibility is 0.36 (MJ/kg).

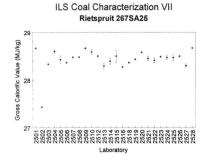
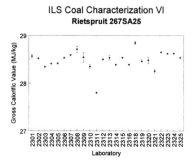

Figure 5 *Results for gross calorific value (MJ/kg) in ILS VII (left) and ILS VI (right) on Rietspruit*

For the other parameters being part of this checking, similar agreement was obtained. It is important to mention that the evaluation of the results of ILS Coal Characterisation VII was carried out first, before a comparison with round VI was made. The software used for evaluation was not designed for working with more than one sample at a time.

11 CERTIFICATION

The results obtained for Rietspruit from ILS Coal Characterisation VI have been summarised in table 1. Recalling from ISO 5725-2:1994, the repeatability (r) is defined as 2.8 times the repeatability standard deviation, and the reproducibility (R) is defined as 2.8 times the reproducibility standard deviation. p denotes the number of laboratories reporting the parameter, excluding those reporting outliers.

Table 1 *Performance Characteristics for Rietspruit (267SA25, ILS VI)*

No	Parameter	m	r	R	p
1	*Inherent moisture (weight %)*	*4.48*	*0.20*	*0.77*	*24*
2	Volatile matter (weight %, dry)	27.14	0.42	1.20	22
3	Ash (weight %, dry)	13.48	0.18	0.43	24
4	Gross calorific value (MJ/kg, dry)	28.51	0.12	0.36	23
5	Carbon (weight %, dry)	72.19	0.47	1.82	12
6	Hydrogen (weight %, dry)	4.06	0.16	0.96	14
7	Sulphur (weight %, dry)	0.58	0.05	0.13	22
8	Nitrogen (weight %, dry)	1.61	0.09	0.57	14
9	Chlorine (weight %, dry)	0.01	0.01	0.02	10
10	SiO_2 (weight % in ash)	45.04	1.08	4.32	13
11	Al_2O_3 (weight % in ash)	30.52	1.10	2.42	12
12	Fe_2O_3 (weight % in ash)	4.77	0.17	0.77	12

No	Parameter	m	r	R	p
13	TiO_2 (weight % in ash)	1.66	0.17	0.34	12
14	MnO_2 (weight % in ash)	0.06	0.01	0.08	12
15	Na_2O (weight % in ash)	0.13	0.05	0.10	11
16	K_2O (weight % in ash)	0.60	0.10	0.19	12
17	CaO (weight % in ash)	6.96	0.30	1.62	12
18	MgO (weight % in ash)	1.67	0.10	0.26	12
19	P_2O_5 (weight % in ash)	1.56	0.51	0.81	12
20	SO_3 (weight % in ash)	4.94	0.39	2.14	11

The value of inherent moisture is indicate, as its actual value depends on the laboratory climate. The certification of the materials took place on the level of the reproducibility. Dividing this value by 2.8 and multiplying it by $k = 2$ yielded the certified uncertainty.

Within the framework of method-defined parameters (inherent moisture, volatile matter, ash content, carbon, hydrogen, nitrogen, sulphur and chlorine content) it is not unreasonable to assume that the involvement of about 25 laboratories will sufficiently "randomise" most effects systematic to each of the laboratories, that the reproducibility standard deviation is a good estimator for the combined standard uncertainty of the material.

This assumption is as good as the assumptions underlying the statistical protocol (ISO 5725-2) are met. The statistics of ISO 5725-2 are based on an ANOVA-approach and work well under the following conditions

1. a single method is involved, or in case multiple methods are used, the repeatability standard deviation may be expected to be the same for all laboratories
2. the number of reported results by the laboratories is about the same for all laboratories

The latter condition is not unimportant. Recent results in interlaboratory studies [16] have clearly demonstrated the problems in establishing a consensus value (grand mean) from unbalanced data. Taking the laboratory averages in those cases as starting point, is not really a solution either: a laboratory average based on for example 20 observations is something different from a laboratory average on only 2 observations. If the participating laboratories do not follow the protocol, the CRM-producer has a potential problem.

For the Rietspruit coal, all assumptions have been verified and were met. Certification took place on the basis of the following confidence interval

$$CI = m \pm k \cdot s_R$$

where s_R is the reproducibility standard deviation, m the grand mean, and a coverage factor $k = 2$. Following from the ISO 5725-2, this CI applies to every single measurement, rather than to the uncertainty on m.

12 USE OF THE PREPARED CRMS

From the analysis made above, the meaning of the conference interval becomes apparent: it is the interval needed for setting up a Sheward-chart. The CRM is available in units of about 60 g, which allows to use it in a pretty short period of time on all parameters relevant to the laboratory.

A possible lay-out of a Sheward chart is given in figure 6.

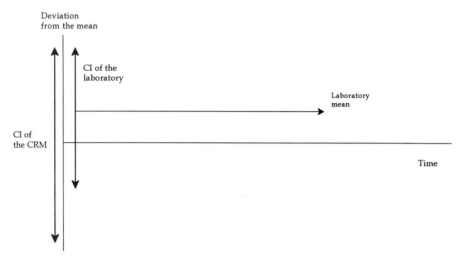

Figure 6 *Sketch of a Sheward-chart*

Given the fact, that a Sheward chart works with two levels, in fact 2s and 3s, the CI for any measurement is given by

$$CI = m \pm 3 \cdot s_R$$

This is called in figure 6 "CI of the CRM". Any valid measurement should, with a level of confidence of more than 99 % (normality of measurement result assumed) be within this CI. However, a laboratory is operating under repeatability conditions, or under "within-laboratory" reproducibility conditions. An estimator for this level is found in s_r, although it does not really account for day-to-day variation. A laboratory may quantify this source of uncertainty, and add it to s_r by means of summation of squared uncertainties and taking the square root. Suppose that the laboratory did not find any significant influence, it can use the limits

$$CI = m_{lab} \pm 3 \cdot s_r$$

as an additional requirement for working under repeatability conditions. Figure 6 is in a sense an exaggeration: this "repeatability CI" should be within the CI of the CRM, and figure 6 represents the extreme case where the top limits of both intervals are at the same value. m_{lab} is the long-term average (which usually fluctuates) of the laboratory, which may tend to m due to improvement of the method or better calibration.

Acknowledgement

The European Coal and Steel Community (ECSC) is acknowledged for their financial support of this work done under contract number ECSC 7220/EC-036.

References

[1] Veen A.M.H. van der, Broos A.J.M., "Preparation and characterisation of coal samples and maceral concentrates for studies on gasification and combustion reactivity of coals in combined cycle processes", Draft final report, ECSC 7220/EC-036, Eygelshoven (NL), 1996

[2] International Organization for Standardization (1989) "ISO Guide 35:1989 - Certification of reference materials - General and statistical principles", second edition, ISO Geneva

[3] Veen A.M.H. van der, "ILS Coal Characterisation I", Evaluation report, NMi Van Swinden Laboratorium B.V., Eygelshoven (NL), 1994

[4] Veen A.M.H. van der, Broos A.J.M., "ILS Coal Characterisation II", Evaluation report, NMi Van Swinden Laboratorium B.V., Eygelshoven (NL), 1994

[5] Veen A.M.H. van der, Broos A.J.M., "ILS Coal Characterisation III", Evaluation report, NMi Van Swinden Laboratorium B.V., Eygelshoven (NL), 1995

[6] Veen A.M.H. van der, Broos A.J.M., "ILS Coal Characterisation IV", Evaluation report, NMi Van Swinden Laboratorium B.V., Eygelshoven (NL), 1995

[7] Veen A.M.H. van der, Broos A.J.M., "ILS Coal Characterisation V", Evaluation report, NMi Van Swinden Laboratorium B.V., Eygelshoven (NL), 1995

[8] Veen A.M.H. van der, Broos A.J.M., "ILS Coal Characterisation VI", Evaluation report, NMi Van Swinden Laboratorium B.V., Eygelshoven (NL), 1996

[9] Veen A.M.H. van der, Broos A.J.M., "ILS Coal Characterisation VII", Evaluation report, NMi Van Swinden Laboratorium B.V., Eygelshoven (NL), 1996

[10] Veen A.M.H. van der, Broos A.J.M., "ILS Coal Characterisation VIII", Evaluation report, NMi Van Swinden Laboratorium B.V., Eygelshoven (NL), 1996

[11] International Organization for Standardization, "ISO 5725-1:1994 Accuracy (trueness and precision) of measurement methods and results - Part 1: General principles and definition", Statistical methods for quality control Vol. 2 (1994), pp. 9-29

[12] International Organization for Standardization, "ISO 5725-2:1994 Accuracy (trueness and precision) of measurement methods and results - Part 2: Basic method for the determination of repeatability and reproducibility of a standard measurement method", Statistical methods for quality control, Vol. 2 (1994), pp. 30-74

[13]   International Organization for Standardization, "ISO/IEC Guide 43-1: voting draft 1996, Proficiency testing by interlaboratory comparisons - Part 1 : Development and operation of proficiency testing schemes"

[14]   BIPM, IEC, IFCC, ISO, IUPAC, IUPAP, OIML (1993) "Guide to the expression of uncertainty in measurement", first edition, ISO Geneva, 1993

[15]   Van der Veen A.M.H., Broos A.J.M., Alink A., "Relationship between performance characteristics obtained from an interlaboratory study programme and combined measurement uncertainty: a case study", Accreditation and Quality Assurance **3** (1998), pp 462-467

[16]   Van der Veen A.M.H., unpublished results

Gaseous Reference Materials – A Challenge in Preparation and Stability

M. Hedrich and H.-J. Heine

FEDERAL INSTITUTE OF MATERIALS RESEARCH AND TESTING (BAM), D-12200 BERLIN, GERMANY

1 INTRODUCTION

In the early days of material's testing it became obvious that comparable test results can only be achieved by using the same standards. Along with the U.S. National Bureau of Standards (NBS, founded 1901; now NIST) the Royal Bureau of Material's Testing (founded 1904; now BAM) in Berlin, Germany, was one of the first institutes to develop reference methods and materials. The precise determination of carbon in steel was the most urgent analytical problem to be solved at that time - other elements followed. Iron and steel still are the predominant matrices for reference materials (RMs) produced by BAM and its cooperation partners followed by non-ferrous metals like Al, Cu, Pb and their alloys. This variety was further enriched by organic substances, porous materials, soils and others in recent years. German legislation passed an act for the nation-wide control of automobile exhaust emission thus initiating the production of high quality gaseous RMs at BAM in 1978.

2 GASEOUS REFERENCE MATERIALS

Gaseous reference materials are gas mixtures of at least two components of well-known fractions with stated uncertainties. As they are primarily being used for the calibration of instruments or measuring methods gaseous RMs are often referred to as calibration gases. Throughout this work both terms will be used synonymously.

According to their chemical nature and the different fields of application accuracy demands vary with respect to their uncertainty ranges. Not only in the area defined by legislation, gaseous RMs follow a certain hierarchy. Following the traceability chain we distinguish between calibration gases of first, second and third order with increasing uncertainty ranges of the components. First order calibration gases are also called primary reference gas mixtures (PRGMs) and meet the highest metrological quality standards. All gaseous RMs, however, have to fulfill the following minimum requirements.

- Qualitative and quantitative composition has to be known and assured for. The appropriate VDI-guidelines[4-17] are a valid tool to verify the composition.
- It has to be made sure that no chemical reactions among the components take place.
- The gaseous RM may not contain any component incompatible with the measuring system to be tested or calibrated.
- Condensation of any component has to be prevented within the temperature (and pressure) range the gaseous RM has been designed for.
- The composition has to remain unchanged during transfer to the measuring system.
- Losses by absorption or adsorption on the inner surface of the container have to be avoided.
- The composition has to be stable throughout the duration of use (validity) of the gaseous RM.

Gaseous RMs which are designed as primary reference gas mixtures (PRGMs) have to meet the demands of the highest quality level. Special requirements concerning the method of preparation and stability are requested in this case.

There is a choice of a number of procedures for the preparation of a gaseous RM. But prior to this step the following questions have to be answered.

- What is the gaseous RM needed for?
- What does it have to be traceable to?
- What composition is it supposed to have?
- Which measuring range will it be valid for?
- Which safety aspects have to be considered during the preparation process (e.g. combustible components in connection with oxygen)?
- Which ranges of uncertainty are tolerable for the individual components?
- How much of the gaseous RM is needed?
- When will it be needed?
- Will there be a reasonable price?

From the answers to these questions one can decide how the gaseous RM has to be prepared or if it can more favorably be purchased by one of the renowned gas producing companies.

3 PROCEDURES

Four groups of techniques are commonly used as methods of preparation. We distinguish roughly between volumetric, manometric and gravimetric procedures as well as procedures using the comparison method. Table 1 gives an overview of the contents of VDI guideline 3490[1-17] including the application (concentration) ranges that can be performed by the individual methods.

Table 1 *VDI Guideline 3490 „Measuring of Gases - Calibration Gases "[1-17]*

Part	Short Form of Title	T	E	W	I
1	Terms and Explanations	Concentr. Ranges T: Technical E: Emission W: Working I: Immission			
2	Preparation Procedures - Overview				
3	Calibration Gases - Transfer				
4	Gravimetric Preparation	T	E	*	*
5	Determination of Composition by Comparison Methods	T	E	W	I
6	Preparation with Gas Mixing Pumps	T	E	*	*
7	Periodic Injection		E	W	*
8	Continuous Injection		E	W	*
9	Permeation into the Matrix Gas			W	I
10	Mixing Streams of Volume - Capillary Dosage			W	I
11	Volumetric Preparation with Plastic Bag	T	E	W	
12	Manometric preparation	T	*	*	*
13	Preparation with Saturation Methods		E	W	
14	Volumetric Preparation in Glass Containers		E	W	I
15E	Direct Determination of Composition - Gas Density Scales	T	E		
16	Preparation by Blending and Mixing	T	E	W	*
17	Thermal Mass Stream Control	T	E	*	

* Calibration gases for these concentration ranges have to be prepared via premixtures.

3.1 Volumetric Preparation

There are two different approaches to apply volumetric preparation procedures - a static and a dynamic one.

3.1.1 Static Procedures[11,14]   Definite volumes of matrix gas and desired component(s) are introduced into a closed vessel and mixed thoroughly. Useful vessels are glass containers, plastic bags with metal lining and other gas-tight containers provided that the inner surfaces do not react with any of the components. Following this way it is in most cases only possible to prepare small amounts of calibration gases in the range of atmospheric pressure.

3.1.2 Dynamic Procedures[6-10,13,16] In this case known (dosable) streams of volume of matrix gas and component are being mixed. As a result a continuous stream of the desired calibration gas is offered. Usually only gas mixtures of two components are available by this way of preparation. To prove proper function of the instrumentation (nominal composition) the actual composition of the furnished gas mixture has to be tested regularly following VDI guidelines. This is common to all preparation devices for calibration gases (calibration gas generators) operating by the principle of a dynamic volumetric procedure. Calibration gas generators have to cope with the drawback of changing gas composition due to a drift during the production process. The risk of erroneous results increases if they cannot be kept running during the time of transportation to the location of the instrument to be calibrated.

Volumetric procedures - static as well as dynamic ones - provide directly volumetric concentrations of the components at atmospheric pressure. Because the resulting gas mixture is not an ideal gas, this has to be considered in the calculation of mass concentrations and/or molar fractions. Standards DIN 1871[18] and DIN 51896 Part 1[19] provide adequate formulae for non-ideal behaviour.

3.2 Manometric Preparation

Manometric procedures are (static) procedures with the potential to produce economically large amounts (volumes) of calibration gas in high pressure cylinders. To achieve this the components and the matrix gas are introduced one by one into the prepared cylinder with the pressure being measured after each dosage. A necessary condition is a constant temperature within the cylinder.

The connection between the partial pressures of the components of a gas mixture (p_i) and its overall pressure (p) is expressed by Dalton's law:

$$p = \sum_i p_i$$

Unfortunately, this simple equation holds true for mixtures of ideal gases only. Especially under high pressure „real" gases exhibit a considerably different behaviour. More elaborate calculation procedures to determine the individual molar fractions from the originally measured (partial) pressures are given in VDI guideline 3490[12]. They take into account procedures of Dalton, Amagat and Kay.

3.3 Gravimetric Preparation

Like the manometric preparation the gravimetric method is a static procedure during which the components and the matrix gas are introduced one after the other into a prepared cylinder. Yet in this case the mass of the cylinder (including its contents) is determined before and after each introduction. The mass of the component is given by the difference of the two weighings and the molar fractions can be calculated directly from this information.

This method can be performed in two ways:

- Determining the mass of the cylinder in an evacuated chamber.
  This procedure is mentioned in VDI guideline 3490[4] but it is of little practical use.
- Determining the mass of the cylinder at atmospheric pressure[20].
  In this case buoyancy and other gravimetric effects influencing the outer surface of the cylinder have to be taken into account.

3.4 Comparison Method[5,21]

In contrast to the three procedures mentioned above (volumetric, manometric and gravimetric procedure) during the comparison method no new gas mixture is produced. In this case the analytical determination of the components of an appropriate gas mixture at hand is used to characterize it as a calibration gas. If such an analytically verified gas mixture can be considered as a gaseous reference material, however, must be scrutinized, because in this case it has to fulfill the minimum demands for gaseous RMs (mentioned in clause 2) as well.

4 GRAVIMETRIC PREPARATION - AN EXAMPLE

Of all methods used to prepare calibration gases the gravimetric preparation has an outstanding position. The uncertainties that can be reached are very small and - how else could it be - the know-how, equipment, time, work and money involved in order to perform this are extensive. One of the duties of our laboratory is the official certification of second order calibration gases which is achieved by gas chromatographic analysis (comparison method[21])using a GC calibrated with PRGMs (first order calibration gases). The preparation of such a primary reference gas mixture will be described in the following clauses.

4.1 Purpose and Planning

In the example chosen here a gas mixture has to be certified as a second order calibration gas. Table 2 shows the approximate composition of the gas mixture.

Table 2 *Approximate Composition of an Exhaust Emission Gas Mixture*

Component	Formula	Molar Fraction / %
Propane	C_3H_8	0.02
Carbonmonoxide	CO	0.5
Carbondioxide	CO_2	6.0
Nitrogen	N_2	Matrix

This calibration gas must match the requirements[22] for the testing of instruments able to measure the automobile exhaust emission composition. One of these requirements is that the relative uncertainty of the certified result of each component may not exceed 1%. This pretentious goal can be hit provided that

- the relative uncertainty of each component of the PRGM may not exceed 0.1% and
- the GC (certification) determination is performed using the bracketting technique with a concentration scale of about ± 10%.

The latter implies that for every second order calibration gas to be certified two PRGMs with the same components have to be at hand (or to be prepared) bracketting tightly enough the gas mixture to be certified.

The first condition demands the highest obtainable quality level concerning preparation (gravimetry) and stability of the composition. This is a characteristic of the first link of the traceability chain, a PRGM. Yet such a low uncertainty cannot be achieved in one step. With an assessed uncertainty of 10 mg per weighing process two premixtures are necessary to keep the overall uncertainty of the components of the final mixture still below 0.1%.

One of PRGMs needed (L3, the one with the concentrations about 10% lower than the calibration gas to be certified) reveals its principal way of formation in the preparation scheme of figure 1.

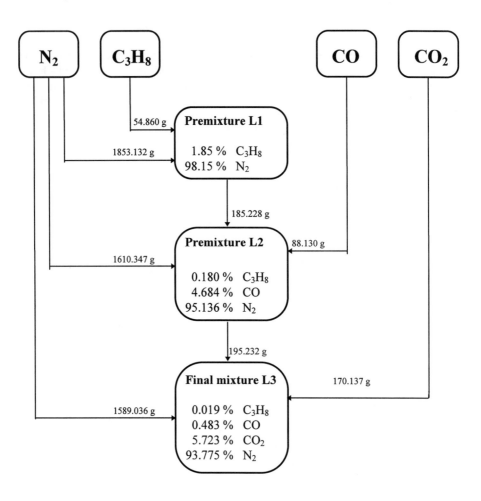

Figure 1 *Preparation scheme of a gravimetrically prepared calibration gas*

The PRGM with higher concentrations - needed as the second partner for calibration with the bracketting technique - has to be prepared in the same manner (H1 to H3) . Not only the final mixtures but also all premixtures share the accuracy level of a first order calibration gas.

4.2 Cylinder Preparation

Unlike most other RMs the gaseous ones are under pressure in an intense contact with the inner surfaces of their containers. So before starting to fill them the cylinders have to be made ready for use. It has to be made sure that they are dust-free, leaktight and do not react chemically or physically with any of the components in measurable quantities.

In our example CO is one of the components. With its ability to form metal carbonyles and its special preference for iron as a reaction partner this gas mixture cannot be kept in an iron containing vessel as a stable calibration gas. Cylinders made of aluminium are the first choice for the storage of gaseous RMs containing carbonmonoxide.

4.2.1 Cleaning   Freshly manufactured cylinders are inspected visually on the inside to detect any irregularities or dirt. If they are dusty they are turned upside down and the dust is blown out with a stream of dry nitrogen. Then the valve is connected to the cylinder. Heating to 70 °C and evacuating are the next (or - in case of reused cylinders - the first) steps. Afterwards the cylinder is filled with dry nitrogen to a pressure of 5 bars and heated to 70 °C again before it is evacuated another time. A heating period lasts 8 hours as a minimum. This cycle has to be repeated at least three times.

4.2.2 Stability Testing   Next the behaviour of the cylinder's inner surface towards the components of the calibration gas is checked. To do so the cylinder is filled with a gas mixture (conditioning mixture) that contains the same components in similar concentrations (maximum of deviation: 20% rel.) as the future calibration gas. After the filling procedure the conditioning mixture is homogenized and its composition analysed. During a storage period of at least 4 weeks the cylinder is heated once to 70 °C for about 8 hours. At the end of the storage period the composition of the conditioning mixture is analysed again. If deviations occur compared to the result of the first analysis the cylinder has to be put aside and the whole procedure must be repeated with another container (of a different material and/or history).

4.3 Time Table

After the cylinder preparation (six are needed in this example) is finished satisfactorily, the actual tasks for the production of the gaseous RMs will begin. The time involved and the steps that have to be taken in the gravimetric preparation of a pair of calibration gases are summarized in table 3.

Table 3 *Preparation Steps for a Pair of Primary Reference Gas Mixtures (Cylinder Preparation not Included)*

Day	Task
1	Evacuate two cylinders for the premixtures L1 and H1.
2	First weighing of evacuated cylinders.
3	Repeat weighing to test vacuum stability (leak tightness). Fill first component (propane) into evacuated cylinders.
4	Weigh the cylinders to determine the amount of propane. Add second component (nitrogen) into cylinders.
5	Weigh the cylinders to determine the amount of nitrogen. Calculate molar composition of the gas mixtures L1 and H1.

Day	Task
	Heat cylinders to 70 °C.
6+7	Homogenize (8 to 16 h) and heat gas mixtures to 70 °C (8 to 16 h).
7	Homogenize gas mixtures L1 and H1. Evacuate two cylinders for the premixtures L2 and H2.
8	First weighing of evacuated cylinders. Analytical validation of premixtures L1 and H1.
9	Repeat weighing to test vacuum stability (leak tightness). Fill first component (L1 or H1 resp.) into evacuated cylinders.
10	Weigh the cylinders to determine the amount of L1 or H1 respectively. Add second component (carbonmonoxide) into cylinders.
11	Weigh the cylinders to determine the amount of carbonmonoxide. Add third component (nitrogen) into cylinders.
12	Weigh the cylinders to determine the amount of nitrogen. Calculate molar composition of the gas mixtures L2 and H2. Heat cylinders to 70 °C.
13+14	Homogenize (8 to 16 h) and heat gas mixtures to 70 °C (8 to 16 h).
14	Homogenize gas mixtures L2 and H2. Evacuate two cylinders for the final gas mixtures L3 and H3.
15	First weighing of evacuated cylinders. Analytical validation of premixtures L2 and H2.
16	Repeat weighing to test vacuum stability (leak tightness). Fill first component (L2 or H2, respectively) into evacuated cylinders.
17	Weigh the cylinders to determine the amount of L2 or H2, respectively. Add second component (carbondioxide) into cylinders.
18	Weigh the cylinders to determine the amount of carbondioxide. Add third component (nitrogen) into cylinders.
19	Weigh the cylinders to determine the amount of nitrogen. Calculate molar composition of the gas mixtures L3 and H3. Heat cylinders to 70 °C.
20+21	Homogenize (8 to 16 h) and heat gas mixtures to 70 °C (8 to 16 h).
21	Homogenize gas mixtures L3 and H3.
22	Validate L3 and H3 analytically by testing compatibility with calibration gases of similar composition (members of the same or a related family).

Provided that no difficulties occur and every step is a success the production of two new three-component PRGMs including cylinder preparation takes a little more than two months. The certification analysis of the calibration gas whose composition is only known approximately (table 2) can start after that. Due to the GC being calibrated with gravimetrically prepared PRGMs (definitive method) the certification will result in a second order calibration gas linked directly to them. As the PRGMs are of similar concentration nonlinearities of the gas chromatographic response only have a small influence. Unfortunately the analytical uncertainty exceeds that of the gravimetric preparation by an order of magnitude. So the relative uncertainties of the analytically certified values will remain below 1%.

4.4 Stability

A RM can only be used correctly as long as its property values (molar fractions of components in this case) are reliable and remain unchanged. The nature of the substances involved is responsible for the chemical stability. This is given as long as no reactions among the components take place and reaction or absorption phenomena (chemisorption) between gas and container material are excluded also.

Physical stability is primarily controlled by temperature and pressure. So storage conditions and modes of use have to be respected that avoid condensation (inhomogeneity) or adsorption (physisorption) of any of the components to cylinder walls, outlet valves or the transfer system.

Any change of the property values with time is also intolerable. A gas mixture will lose its status as a RM if alterations occur concerning its composition during storage or after repeated handling.

To control and guarantee stability as far as possible, PRGMs as a rule do not leave our house and undergo regularly a stability testing program involving all members of their own and selected ones of related families. Second order calibration gases are furnished with a date of validity (expiry date) on their certificates.

5 SOURCES OF UNCERTAINTY[20]

Manifold sources take part in influencing the individual uncertainties of the molar fractions of the components of a PRGM. The most prominent ones are:

- weighing uncertainty of the scales;
- nonlinearity, wrong zero calibration, thermal and time-influenced drift of the scales;
- place where the cylinder is located on the scales;
- uncertainty and buoyancy of the mass standards;
- abrasion of matter from the cylinder (container outside wall, windings, valve);
- dust;
- absorption or desorption of water on the outer wall of the cylinder;
- buoyancy change of the cylinder as a consequence of change in volume, temperature change, volume change by pressure of the filling and changes of air density (temperature, pressure and humidity);
- uncertainties of gas composition;
- uncertainties of molar masses;
- inhomogeneities;
- leakage of the cylinder and in the transfer system (inlet of air into evacuated cylinder, gas loss after filling);
- change of gas composition in the transfer system;
- absorption and adsorption at the inner surface of the cylinder;
- reactions between the components.

The combined uncertainties of the PRGM and of the analytical method used determine the overall uncertainty of the second order calibration gas prepared with the comparison method.

6 USE OF GASEOUS REFERENCE MATERIALS

The benefit one can draw from a calibration gas is directly connected to its use. Two aspects that need consideration are how to deal with it properly and what to use it for.

6.1 Proper Handling

Gaseous RMs require a proper way of being handled during storage, transportation and analysis on the producer's as well as on the user's side.

For a second order calibration gas the storage time begins when the user receives the cylinder and ends when either the minimum utilization pressure is reached or the end of the stability time period. Both limitations are stated in the certificate. The lower limit of the storage temperature is related to the dew point of the gas mixture at the maximum filling pressure and is specified in the certificate as well. Cylinder valves have to be kept clean and protected from humidity to prevent corrosion. The use of cylinder valve caps is required.

During transport, which is only allowed on the road, by boat or by train, the temperature of the cylinder must exceed the lower limit of the storage temperature.

Special care has to be taken if gas is withdrawn from the cylinder (for analysis) concerning the pressure difference to the transfer system, the Joule Thompson cooling effect and the minimum temperature with respect to the dew point to avoid unmixing of components. The materials that come into contact with the gas mixture have to be clean, impermeable, of negligible sorption capability and chemically inert. Highly recommended materials are stainless steel and aluminium. The transfer system has to be leak tight and purged (rinsed) effectively with the calibration gas prior to analysis.

6.2 Fields of Application

As any other RM, a gaseous reference material can serve several purposes. Most commonly it is used for the calibration of measuring instruments like gas chromatographs, calorimeters, CO_2 measuring devices, combustion or extraction furnaces for the determination of gases in solids, etc.

After the development of a new analytical procedure, RMs are needed to evaluate the new measurement method and compare it to others. Only if we know what we expect to find can we judge whether the way that leads to the result is a suitable one or not.

Quality assurance in an analytical laboratory calls for RMs as well. In this case they are used as analytical control samples passing the procedure the same way as the specimen whose composition is to be quantified. By finding the certified results within their ranges of uncertainty one can assume that the analytical process is under control and the results are trustworthy.

Moreover, a RM can serve as a tool to assess the performance of testing laboratories in interlaboratory comparisons or accreditation practice. Keeping the various fields of application in mind, RMs are equally important for instrument manufacturers, research and development laboratories and quality control in industrial production.

7 OUTLOOK

The hierarchy of the gaseous RMs (first, second and third order calibration gases with each of them being linked to the one next order) is a tool to provide vertical traceability within Germany. Other nations maintain similar systems. But how do the PRGMs of the national institutes all over the globe compare with each other? To establish this kind of horizontal traceability, BIPM organizes international key intercomparisons via its consultative committee like the one dealing with automobile emission gases initiated in the second half of 1998.
Others will follow.

ABBREVIATIONS

BAM	Bundesanstalt für Materialforschung und -prüfung
BIPM	Bureau Internationale de Poids et Mesures
DIN	Deutsches Institut für Normung
GC	Gas chromatograph(y)
ISO	International Standardisation Organisation
NBS	National Bureau of Standards
NIST	National Institute of Standards and Testing
PRGM	Primary reference gas mixture (first order calibration gas)
PTB	Physikalisch-technische Bundesanstalt
RM	Reference material
VDI	Verein deutscher Ingenieure

REFERENCES

1. VDI 3490, Blatt 1, Beuth Verlag, 1980
2. VDI 3490, Blatt 2, Beuth Verlag, 1980
3. VDI 3490, Blatt 3, Beuth Verlag, 1980
4. VDI 3490, Blatt 4, Beuth Verlag, 1980
5. VDI 3490, Blatt 5, Beuth Verlag, 1980
6. VDI 3490, Blatt 6, Beuth Verlag, 1988
7. VDI 3490, Blatt 7, Beuth Verlag, 1980
8. VDI 3490, Blatt 8, Beuth Verlag, 1981
9. VDI 3490, Blatt 9, Beuth Verlag, 1980
10. VDI 3490, Blatt 10, Beuth Verlag, 1981
11. VDI 3490, Blatt 11, Beuth Verlag, 1980
12. VDI 3490, Blatt 12, Beuth Verlag, 1988
13. VDI 3490, Blatt 13, Beuth Verlag, 1992
14. VDI 3490, Blatt 14, Beuth Verlag, 1994
15. VDI 3490, Blatt 15 E, Beuth Verlag, 1985
16. VDI 3490, Blatt 16, Beuth Verlag, 1994
17. VDI 3490, Blatt 17, Beuth Verlag, 1998
18. DIN 1871, „Gasförmige Brennstoffe und sonstige Gase - Dichte und andere volumetrische Größen", Beuth Verlag, 1998
19. DIN 51896, Teil 1, „Gasanalyse - Zusammensetzungsgrößen, Realgasfaktor - Grundlagen", Beuth Verlag, 1998

20. ISO/DIS 6142, „Gas analysis - Preparation of calibration gas mixtures - Gravimetric methods", 1998
21. ISO/DIS 6143, „Gas analysis - Determination of composition of calibration gas mixtures - Comparison methods", 1998
22. PTB-A 18.10, „Meßgeräte im Straßenverkehr - Abgasmeßgeräte für Fremdzündungsmotoren", 1992

Uses of Matrix Reference Materials

Henry F. Steger

CANADIAN CERTIFIED REFERENCE MATERIALS PROJECT, MINING AND MINERAL
SCIENCES LABORATORIES, NATURAL RESOURCES CANADA, 555 BOOTH STREET,
OTTAWA, ONTARIO, K1A 0GI, CANADA

1 INTRODUCTION

"Compositional" or matrix reference materials such as the certified reference metals, alloys, rocks, ores, concentrates, etc., are in common use in analytical laboratories. The simplest and possibly best definition of a matrix reference material is that it is a "natural" substance more representative of laboratory samples that has been chemically characterized for one or more elements, constituents, etc. with a known uncertainty. "Natural" materials are more complex, chemically, compositionally and structurally, than primary reference standards, which are often pure chemicals. On the other hand, the chemical characterization of "natural" materials usually has a greater level of uncertainty and is less straightforward in terms of traceability. It is also true to say that the majority of available matrix reference materials have been certified through interlaboratory measurement programs. This presentation touches on some of the factors of matrix reference materials that a user should take into careful consideration in their use.

2 PROPERTIES OF MATRIX REFERENCE MATERIALS

It is in general true that an analytical laboratory equates the "use" of reference materials with "the agreement between its measured value and the certified value". In fact however, the use of a matrix reference material actually begins in the selection of an appropriate reference material.

A matrix reference material has several properties that can influence a user to choose it for a specific purpose. The relative importance of these properties can vary depending on the intended use, i.e., to assess performance or to calibrate an instrument.

Some of the important properties of matrix reference materials are obvious. For example:
1. certified elements/constituents;
2. matrix, i.e., chemical and/or mineralogical composition, and
3. concentration of the certified elements.

(1) This point seems redundant. Surely a user would not use a reference material certified for, e.g., zinc, to assess performance for lead! However it does occur that a user cannot find a suitable reference material and in desperation, or possibly ignorance, uses an element with an uncertified value in an available reference material to assess performance. My organization, the Canadian Certified Reference Materials Project (CCRMP), received a complaint last year from a client who stated that his use of the value for Al in reference material MP-1a clearly demonstrated that it was too high by ~20%. This occurred despite the fact that CCRMP had stated that the value for Al was for general compositional information only! The value was based on duplicate analyses performed in a "quick and dirty" manner.

(2) The main advantage that matrix reference materials offer over primary reference standards is that they provide a better match to sample composition. They are a tool for minimizing "matrix" or inter-element effects or identifying shortcomings in the analytical method such as incomplete decomposition, etc. The knowledgeable user strives to select matrix reference materials having certified elements in a matrix as closely resembling the samples to be analyzed as possible.

Conversely, the user should be aware that the matrix reference materials may contain one or more elements which are not present in the samples but which, if at a sufficiently high concentration, could give rise to matrix or inter-element effects so that there is in effect no true matrix matching.

(3) In selecting a matrix reference material to match the matrix of his samples, the user should consider the magnitude of the certified value(s). There is a potential danger that, in preparing the solution of the reference material at the appropriate concentration for the element(s) of concern, the overall solution composition could differ significantly from that of the samples to be analyzed. This could apply equally to a high certified value which would require appreciable dilution of the prepared solution or a low certified value which would lead to a solution of higher concentration in matrix elements than found in the samples.

Other properties of matrix reference materials to be considered by the user are, inter alia:

4. the period of validity of the certified values;
5. prescribed conditions for storage;
6. instructions for use such as:
   - minimum test portion amount,
   - use "as is" or "dry at 100°C, etc.;
7. certified values may be specific method dependent;
8. magnitude of the uncertainty of the certified values; and
9. traceability .

(4) Many matrix reference materials are stable indefinitely. Examples are reference materials made from natural stable metal oxides or siliceous material with no or low sulphide content. On the other hand, reference materials made from sulphide ores and concentrates may oxidize and therefore their certified values may have a limited period of validity. The user should not use a reference material for which the date of validity has passed. A revised ISO Guide 31 "Contents of certificates of reference materials" will shortly be issued and it specifies that each reference material be assigned a period of validity. I wonder how many users have not succeeded in attaining a certified value within the stated uncertainty only to discover that the date of validity has passed.

(5) If a reference material has a defined period of validity, its useful lifetime may be unnecessarily shortened if the user does not adhere to the conditions prescribed for storage.

(6) In selecting a matrix reference material, the user should take into account the potential effect of test portion amount. If the test portion amount used to establish the homogeneity of the material was larger than the test portion amount intended by the user, the user could detect inhomogeneity. Similarly, the user could detect inhomogeneity even at the same test portion amount if his method has better precision that that used to establish homogeneity. In both instances, the detected inhomogeneity gives an uncertainty that must be added to the uncertainty of the certified value. The user should be aware that the uncertainty arising from the inhomogeneity could become sufficiently large that using only the statistical parameters of certification is no longer justified - the reference material is unsuitable for the intended purpose. This phenomenon can sometimes be observed in gold ores. An ore established to be homogeneous at one assay tonne can be experimentally shown to be inhomogeneous at smaller test portion amounts. As a result, when only one or two determinations are carried out at the smaller test portion, it is observed that the value determined for gold amounts does not agree with the certified value within the stated uncertainty. A similar problem may arise if the user's method requires a smaller test portion amount than is recommended by the producer of the matrix reference material for use in analysis.

The importance of the role of sample size in the use of reference materials was recognized by ISO/REMCO which subsequently created the Sampling Task Group to address this issue.

(7) The user should be aware that when an element in a matrix reference material has been certified to a level of uncertainty by a specific method, it cannot be assumed that these quantities are necessarily applicable to his method, if different. The user should use this material to assess his performance for an element only where the accuracy and precision of his method are comparable to those of the method used in certification.

(8) The magnitude of the uncertainty of the certified values of a matrix reference material can be an important consideration for the user. Indeed the user may have to consider the uncertainty in both the accuracy and the precision of his method in selecting suitable matrix reference materials. Because the issue is parallel for both accuracy and precision, I will discuss accuracy only.

One possibility occurs when the user's method is of lower uncertainty in accuracy than the mean performance of the methods used in the certification of an element. Here the user in general passes the ISO Guide 33 statistical tests for the assessment of a method for accuracy. In fact however the user is evaluating his performance unrealistically and he should use a different reference material, one that has an uncertainty in the certified value that is commensurate with the capability of his method.

Related to this case is when the user must for some reason, e.g., a business agreement with a client or a need to demonstrate on-going validation of a method due to accreditation status, attain a defined level of accuracy in analysis, i.e. a defined level of uncertainty in accuracy. If the user chooses to prove this level through the use of matrix reference materials, he must select those materials with certified values with an uncertainty that permits this assessment. Of course, all this is meaningless if the user does not have a method capable of this level of accuracy.

A second possibility occurs when the capability of the user's method is of greater uncertainty in accuracy than the mean performance of the methods used in the certification of an element. Here the user could fail the ISO Guide 33 statistical tests for the assessment of a method for accuracy. In fact however the user is evaluating his performance unrealistically and he should use a different reference material, one that has an uncertainty for the certified value that is commensurate with the capability of his method. A good example of this case is the use of a matrix reference material certified for nickel by dimethylglyoxime precipitation, a very accurate and precise method, to assess the performance of a method that has an ICP-AES finish.

In the broader context, the case where the user's method is not capable of matching the accuracy and precision reported for a reference material may be encountered quite frequently when the reference material had been certified by a primary measurement method, i.e., a method of highest metrological quality. An example of common occurrence for this case is the use of reference "waters'. These "waters" have in general been characterized by primary methods by national agencies such as NIST. In both cases, the analyst would indeed be hard pressed to obtain an uncertainty in accuracy and precision at the same level as the certified value.

I will revisit the implications of these two cases after the statistical tests for assessing the accuracy and precision of a method have been discussed below

(9) Traceability is defined as the property of a measurement result whereby it can be related to stated references through an unbroken chain of comparisons, all having stated uncertainties. In ISO Guide 30 "Terms and definition used in connection with reference materials", the definition of a certified reference material requires that the elements of interest be certified by a procedure that establishes traceability to an accurate realization of the unit in which the certified value is expressed.

Establishing traceability for the certified values in a matrix reference material is not necessarily straightforward. The chemical measurements of a certified element could have been made by a broad range of methods which range from primary methods to well-defined methods for which traceability to an internationally agreed measurement scale can be demonstrated. Draft ISO Guide 31 "Contents of certificates, certification reports and labels of reference materials" recommends that a certificate for a reference material contain a clear statement that outlines the principles of the methods used, together with evidence of their validity and the measurement scale to which they are traceable. Where traceability is an important issue, the user should ensure that the reference material selected has the appropriate evidence of traceability.

Another important aspect of "traceability" that an analyst is at times faced with, but which falls outside the realm of this presentation, is the use of "specified" reference materials to demonstrate acceptable accuracy and precision. This can be the case for analyses or measurements related to regulations, clinical applications or even business agreement. In these cases, the user has little choice other than to use the prescribed reference materials or to refuse the work.

3 USE TO ASSESS A METHOD FOR ACCURACY AND PRECISION

The use of matrix reference materials in analytical laboratories has increased appreciably during the past few years, mainly because of the increasing emphasis on and demand from the client for supportable evidence for the quality of the product or service, and also by the mounting acceptance of ISO quality systems for accreditation or certification. Fortunately the level of knowledge about using matrix reference materials to assess proficiency has increased corresponding but there still remain some misunderstandings on how to assess proficiency through statistical tests.

This part of the presentation presents how the user can assess the level of the performance of a method by applying the statistical tests in ISO Guide 33.

3.1 Precision (Repeatability) of a Method

To assess the precision of an analytical method by the analysis of a certified reference material, the average within-laboratory standard deviation of the certified value must be known and the user should have at least duplicate (2) results but preferably three or more. The analytical method is accepted with regard to precision if:

$$(S_{wL}/\sigma_{Rm})^2 \leq (\chi^2_{n;\ 0.95}/[n-1]) \qquad \text{Eq. 1}$$

where:

n	=	the number of replicates (n should be ≥ 2),
S_{wL}	=	the standard deviation of the replicate results,
σ_{Rm}	=	the within-laboratory standard deviation of the certified value, and
$\chi^2_{n;\ 0.95}$	=	0.95-quartile of the χ^2 distribution at $(n-1)$ degrees of freedom.

Note: In practice, this is not necessarily a very demanding test at low n and a laboratory may want or be required to set tighter limits, for example, for Shewhart charts, business agreements, etc..

3.2 Accuracy of a Method

There still remains appreciable misunderstanding that the determined value of an element should lie within the 95% confidence intervals of the certified value for the element in a matrix reference material if the method is acceptable in terms of accuracy. While this is true, it is strictly not correct according to ISO Guide 33 "Uses of certified reference materials" and is also in conflict with the statistics. It should always be remembered that the 95% confidential intervals define the range of values that a certified value would be expected to occur 19 times if the interlaboratory certification program were repeated 20 times under the same conditions. To be noted is that the magnitude of the 95% confidential intervals depends on the number of participating laboratories.

According to ISO Guide 33, the between-laboratories standard deviation of the certified value of a reference material should be known to assess the accuracy of a method. The method should be assessed on the basis of the mean value for two or more replicate results.

3.2.1 Two or More Replicate Analytical Results. The analytical method is accepted with regard to accuracy if:

$$|X_C - X_L| \leq 2\sqrt{\sigma_{Lm}^2 + S_{wL}^2/n} \qquad \text{Eq. 2}$$

X_C	=	the certified value,
σ_{Lm}	=	the between-laboratories standard deviation of the certified value,
X_L	=	mean value of the replicates, and
S_{wL}	=	standard deviation of X_L.

In many cases, $\sigma_{Lm} \gg S_{wL}$, and/or $n > 3$ and Equation 2 can be approximated by:

$$|X_C - X_L| \leq 2\sigma_{Lm} \qquad\qquad \text{Eq. 3}$$

A spread of $\pm 2\ \sigma_{Lm}$ about the certified value contains 95% of the values used to estimate the certified value. About 5% of the accepted values lay outside this spread. Therefore, satisfying Equation 2 indicates that the value found by the method being tested would have been accepted in estimating the certified value of that element. In contrast, there is only a 5% probability that a result for a method that fails Equations 2 or 3 would in fact be acceptable with respect to accuracy.

The 95% confidence intervals (C.I.) can be approximated by:

$$\text{C.I.} \approx \pm\ t_{0.975\ (k-1)} \cdot \sigma_{Lm}\ /\ \sqrt{k} \qquad\qquad \text{Eq. 4}$$

where:

k   = the number of sets of results (laboratories) used to estimate the certified value; and

$t_{0.975\ (n-1)}$ = the 0.975 fractile of the one-sided Student distribution with $(n-1)$ degrees of freedom. Its value can be approximated as "2" at $n \geq 10$.

If we compare Equations 2 or 3 with 4, we see that the C.I. is $\approx \sqrt{k}$ times smaller than the spread allowed by ISO Guide 33. This is why the user should not specifically aim to fall within the 95% C.I. in assessing a method for accuracy. However if the user's method falls within the 95% C.I., so much the better! The following table which shows a comparison of the 95% C.I. and $2\ \sigma_{Lm}$ for some example elements in three matrix reference materials substantiates the $\sqrt{k}$ times relationship.

RM	Element	k	CV	95% C.I.	$2\ \sigma_{Lm}$
SU-1a	Ni	22	1.233%	0.008%	0.0322%
	Cu	24	0.967%	0.005%	0.0230%
	Co	20	0.041%	0.001%	0.004%
CH-3	As	10	143 μg/g	14 μg/g	39.6 μg/g
	S	18	2.82%	0.03%	0.094%
CZN-3	Zn	37	50.92%	0.08%	0.42%
	Ag	31	45 μg/g	2 μg/g	10 μg/g

A relationship that $\sigma_{Lm} \approx 2\ \sigma_{Rm}$ for the certified value is observed for many matrix reference materials, although there is no theoretical or statistical basis for this. When the value of σ_{Lm} is unknown, the user may assume that the method is accepted with regard to accuracy if:

$$|X_C - X_L| \leq 4\sigma_{Rm} \qquad\qquad \text{Eq. 5}$$

Note: The scope of the validity of this assumption is unknown and the user should be aware of the risk in accepting it. For example, the assumption is not valid for low level gold CCRMP reference materials such as GTS-2 and CH-3 for which $\sigma_{Lm} \approx \sigma_{Rm}$.

If neither σ_{Lm} nor σ_{Rm} are known, the laboratory could analyze the reference material at least in triplicate to provide an estimate of S_{wL} which can be used as an approximation for σ_{Rm}. The method would therefore be accepted for accuracy if:

$$|X_C - X_L| \le 4\,S_{wL} \qquad\qquad\qquad \text{Eq. 6}$$

Note: ISO/REMCO does not recommend the use of certified reference materials for which the uncertainty of certified values are not reported. This is usually found only for reference materials issued many years ago.

The test expressed by Equation 5 may however be the "only" approach for a user who uses a reference material certified by a "definitive" method in full knowledge that his method is not capable of matching the accuracy and precision of the reference material.
3.2.2 Single Analytical Result. If there is a single result, the analytical method is accepted with regard to accuracy if:

$$|X_C - X_L| \le 2\sqrt{\sigma_{Lm}^{\,2} + \sigma_{Rm}^{\,2}} \qquad\qquad \text{Eq. 7}$$

As above, if $\sigma_{Lm} \gg \sigma_{Rm}$, the analytical method is accepted with regard to accuracy if the condition in Equation 3 is met:

$$|X_C - X_L| \le 2\sigma_{Lm} \qquad\qquad\qquad \text{Eq. 3}$$

If it is assumed that $\sigma_{Lm} \approx 2\,\sigma_{Rm}$, the analytical method is accepted with regard to accuracy if the condition in Equation 4 is met:

$$|X_C - X_L| \le 4\,\sigma_{Rm} \qquad\qquad\qquad \text{Eq. 5}$$

Similarly when σ_{Lm} of the certified reference material is not known, the analytical method is accepted with regard to accuracy based on a single result according to Equation 4.

3.3.3 Examples. The following are two examples of assessing a method for accuracy and precision using two CCRMP gold ore matrix reference materials, MA-1b and CH-3.

Gold Ore, MA-1b

Au Value				
Certified Value	**95% C.I.**	σ_{Lm}	σ_{Rm}	
17.0 µg/g	0.3 µg/g	0.70 µg/g	0.42 µg/g	$\sigma_{Lm} \approx 2\,\sigma_{Rm}$

	Laboratory A	Laboratory B	
Mean Value, X_L	18.6 μg/g (1)*	17.12 μg/g (5)*	( )* = replicates
St'd Dev., S_{wL}		0.49 μg/g	

		Laboratory A	Laboratory B	
	$\lvert X_C - X_L \rvert$	1.6 μg/g	0.12 μg/g	
Eq. 1	$(S_{wL}/\sigma_{Rm})^2$		1.38	1.38 < 2.78
Eq. 1	$(\chi^2_{n;\ 0.95}/[n-1])$		2.78	
Eq. 2	$2\sqrt{\sigma_{Lm}^2 + S_{wL}^2/n}$		1.47 μg/g	
Eq. 5	$4\,S_{wL}$		1.96 μg/g	
Eq. 6	$2\sqrt{\sigma_{Lm}^2 + \sigma_{Rm}^2}$	1.63 μg/g		
Eq. 4	$4\,\sigma_{Rm}$	1.68 μg/g		

- ♦ Laboratory A and B are accepted with respect to accuracy.
- ♦ Laboratory B is accepted with respect to precision.

Note: Laboratory B also falls within the 95% C.I whereas Laboratory A does not but is
still accepted with regard to accuracy

Gold Ore, CH-3

Au Value				
Certified Value	**95% C.I.**	σ_{Lm}	σ_{Rm}	
1.40 μg/g	0.30 μg/g	0.07 μg/g	0.11 μg/g	$\sigma_{Lm} \approx \sigma_{Rm}$

	Laboratory A	Laboratory B	
Mean Value, X_L	1.15 μg/g (1)*	1.78 μg/g (3)*	( )* = replicates
St'd Dev., S_{wL}		0.082 μg/g	

		Laboratory A	Laboratory B	
	$\lvert X_C - X_L \rvert$	0.25 μg/g	0.38 μg/g	
Eq. 1	$(S_{wL}/\sigma_{Rm})^2$		0.56	0.56 < 3.91
Eq. 1	$(\chi^2_{n;\ 0.95}/[n-1])$		3.91	
Eq. 2	$2\sqrt{\sigma_{Lm}^2 + S_{wL}^2/n}$		0.17 μg/g	
Eq. 5	$4\,S_{wL}$		0.33 μg/g	
Eq. 6	$2\sqrt{\sigma_{Lm}^2 + \sigma_{Rm}^2}$	0.26 μg/g		
Eq. 4	$4\,\sigma_{Rm}$	0.44 μg/g		

- ◆ Laboratory A is accepted with respect to accuracy.
- ◆ Laboratory B is unsatisfactory with respect to accuracy.
- ◆ Laboratory B is accepted with respect to precision.

Note: Neither laboratory falls within the 95% C.I but Laboratory A is still accepted with regard to accuracy

The two examples above point out the risk of using Equation 6 to assess a method for accuracy. In both cases,

$$4 \ S_{wL} \ > \ 2\sqrt{\sigma_{Lm}^{2} + S_{wL}^{2}/n}$$

and therefore the assessment is less stringent.

Let's return to discuss the impact of the relative magnitude of the uncertainty in accuracy of the reference material and the method being assessed. Figure 1 depicts when the uncertainty in accuracy of the reference material is better than that of the method. Figure 1a shows where the method is acceptable with regard to accuracy, i.e., the value determined by the method falls within $\pm 2\sigma_{Lm}$ of the certified value. Figure 1b shows where the value determined by the method is rejected with regard to accuracy, i.e., the value determined by the method does not fall within $\pm 2\sigma_{Lm}$ of the certified value. In fact however, the greater uncertainty of the method renders this test invalid. A good analogy is trying to measure an item several meters in length with a gauge of lesser length and that is marked only in centimeters. The wrong tool has been used! The user should have selected a reference material for which the uncertainty in the certified value such that $2\sigma_{Lm} \approx 4S_w$.

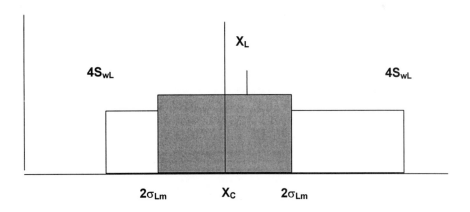

Figure 1a *Case 1: Uncertainty of C.V. < users' method*

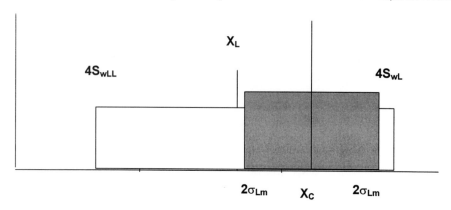

Figure 1b *Case 1: Uncertainty of C.V. < users' method*

Figure 2 depicts when the uncertainty in accuracy of the method is better than that of the reference material. In this instance however Figures 2a and 2b both show that the method is acceptable with regard to accuracy, i.e., the value determined by the method falls within $\pm\ 2\sigma_{Lm}$ of the certified value. The greater uncertainty of the reference material renders this test meaningless. A good analogy is trying to measure an item that is only several centimeters in length with a gauge of greater length and that is marked only in meters. As above, the wrong tool has been used! The user should have selected a reference material for which the uncertainty in the certified value such that $2\sigma_{Lm} \approx 4S_{wL}$.

Gold ore MA-1b can be used to illustrate the above. A quick calculation shows that $2\sigma_{Lm}$ is ~8.2% of the certified value, i.e., MA-1b allows the user to assess accuracy to no better than 8.2%. If the user must demonstrate that the method gives an accuracy to an uncertainty better than ~8.1%, e.g., 5%, he should select a different certified gold ore for which $2\sigma_{Lm} \leq 5\%$. On the other hand, if the user's method gives no better than $\sim \pm 8\%$, MA-1b is a suitable reference material.

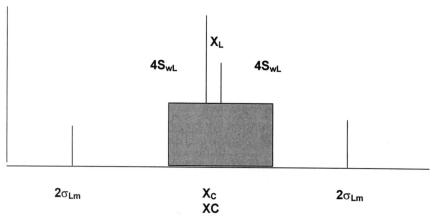

<u>Figure 2a</u> *Case 2: Uncertainty of C.V. > users' method*

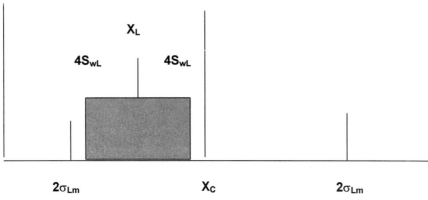

Figure 2b *Case 2: Uncertainty of C.V. > users' method*

4 LONGER TERM ASSESSMENT OF THE ACCURACY OF A METHOD

The statistical tests in ISO Guide 33 provide a snap shot of method performance, i.e., a one time event. What can be done if a laboratory analyzes a given matrix reference material by the same method several times over a period of time? The method can be assessed at each analysis event with the tests in ISO Guide 33 but there might be other important information to be learned on method performance if the analyses were tied together. Figure 3 shows some results from the analyses by Laboratory A of some reference materials certified for arsenic over a 27 month period. Each point shown is a mean of duplicate measurements. Each unit on the abscissa indicates one event of measurement. The points represent the relative accuracy (RA) which is given by the difference between the certified value and the mean of the duplicates expressed as a percentage of the certified value. The solid line is the mean value of the points (MRA). The uncertainty in MRA is the standard deviation expressed as a percentage of the certified value.

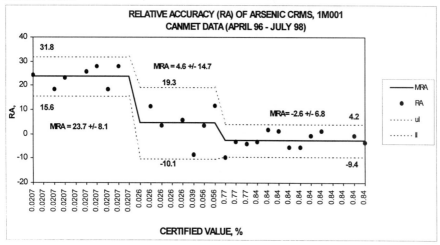

Figure 3 *Longer term results for arsenic in some reference materials*

Similar plots for several elements have pointed out that it is common that a single value of MRA can be calculated for two or more reference materials and/or for two or more methods for a range of concentration. This is the case for arsenic but this observation is obscured by the dominance of data for four reference materials. A closer examination shows that the estimated value for MRA in a "flat" region" can be applied for up to three reference materials.

It is possible to test each mean value for arsenic by the test expressed by Equation 2 or 3 to assess that the mean value is acceptable with respect to accuracy. However, the test can be made more sensitive and, therefore more informative in the longer term, if the determined arsenic values for a given reference material are combined to estimate an overall mean and the associated standard deviation to give:

X_C = the certified value,
σ_{Lm} = the between-laboratories standard deviation of the certified value,
k = the number of sets of results;
n = the number of replicates per set of results;
X_D = overall mean value; and
S_{wD} = average within-set standard deviation.

The method is accepted with respect to accuracy if:

$$|X_C - X_D| \leq [\, 2\sqrt{\sigma_{Lm}^2 + S_{wD}^2/n}\,]/\sqrt{k} \quad \text{Eq. 7}$$

To illustrate this statistical test, the results in Figure 3 are used.

Reference Materials

Reference Material	As Value (%)			
	Cert. Value	95% C.I.	σ_{Lm}	σ_{Rm}
MP-1a	0.84	0.02	0.0349	0.0163
CZN-1	0.026	0.002	0.00445	0.0018
RTS-4	0.0207	0.0044	0.0042	0.0009

Reference Material	As Value (%)	
	Mean Value, X_D	St'd Dev., S_{wD}
MP-1a	0.818 (11)	0.014
CZN-1	0.0272 (5)	0.0007
RTS-4	0.0256 (10)	0.0005

( )* = k

Reference Material	$\lvert X_C - X_D \rvert$	$[\, 2\sqrt{\sigma_{Lm}^2 + S_{wD}^2/n}\,]\,/\sqrt{k}$	Meets (Y/N)
MP-1a	0.022%	0.022%	Y
CZN-1	0.0012%	0.0040%	Y
RTS-4	0.0049%	0.0027%	N

- ◆ Laboratory A is accepted with respect to accuracy for arsenic for reference materials MP-1a and CZN-1.
- ◆ Laboratory A is not accepted with respect to accuracy for arsenic for reference material RTS-4.

A laboratory however may have the option of accepting a certain bias in its method either to a business agreement with a client or to ensure the continued practicality of its operations. Where this is the case, the method is accepted with respect to accuracy if:

$$\lvert X_C - X_D \rvert \;\leq\; b_A + [\, 2\sqrt{\sigma_{Lm}^2 + S_{wD}^2/k}\,]\,/\sqrt{k} \quad \text{Eq. 8}$$

where:

b_A = the absolute value of the acceptable bias.

For the example above, the method used above by Laboratory A to determine arsenic in reference material RST-4 would pass the test for accuracy if a bias of at least ± 0.0022% As were acceptable for operational considerations.

5 USE IN CALIBRATION

There are occasions when the use of matrix reference materials may be preferred for the calibration of instruments. Matrix reference materials may often be the only way of establishing calibration curves for techniques such as X-ray fluorescence by pressed powder pellet. However it should be noted that the user faces the same challenges in selecting matrix reference materials for use in calibration as for the assessment of a method for accuracy and precision, e.g., the user should note that the uncertainty in a certified value is always one of the components in establishing the overall uncertainty of the calibration curve and should select a reference material for both the appropriate certified value and also for an uncertainty of acceptable magnitude.

As discussed above, matrix reference materials provide the opportunity to match the composition of the samples to minimize matrix and inter-element effects and therefore the use of matrix reference materials to establish calibration curves may at times be more effective that the use of calibration standards which are usually an aqueous solution of a simple chemical salt. The ideal case for using matrix reference materials for calibration occurs where the user has a series of reference materials of essentially the same matrix and which provide the desired range of certified values, and where that matrix matches closely that of the samples. This is often applicable to analytical laboratories that support a production company that is restricted to one or only a few types of products, e.g., an iron ore company, cement plant, etc., for which several reference materials of the same matrix are available.

Again as discussed above, the user should be aware that the use of matrix reference materials may introduce matrix and inter-element effects not present in the samples. A problem could therefore arise if the user has to resort to reference materials of significantly different matrices in order to achieve the desired range in certified values. Here there is the potential that one or more reference materials could show matrix or inter-element effects that make them incompatible. To illustrate this, the following matrix reference materials:

Refer. Mat.	Cu Certified Value (%)	Uncertainty (%) $2\sqrt{\sigma_{Lm}^2 + S_{wL}^2/n}$
MP-1a	1.44	
SU-1a	0.967	
KC-1a	0.629	
HV-1	0.53	0.027 ($S_{wL} = 1$)
UM-1	0.42	
CPB-1	0.25	
CZN-1	0.144	

were used to establish response (calibration) curves for an X-ray fluorescence spectrometer (XRF) using pressed powder pellets and for an atomic absorption spectrometer (AAS) after acid dissolution of the reference materials. Figure 4 illustrates the response curve for the X-ray fluorescence spectrometer. It is clear that the response from material HV-1 is not consistent with that from the other reference materials. The response curve shown is the regression line that excludes HV-1. It is easily seen that the range of 0.503 to 0.557% Cu (0.53% ± 0.0274% uncertainty) does not intersect the response curve. In contrast, Figure 5 shows the response curve for the atomic absorption spectrometer. It is clear that HV-1 does not show a response that is inconsistent with that of the other materials.

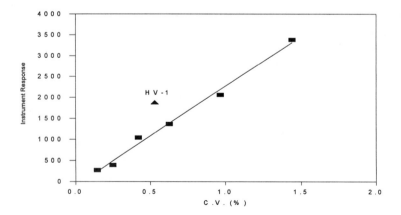

Figure 4 *Cu response curve using matrix reference materials (XRF)*

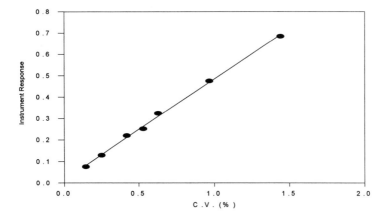

Figure 5 *Cu response curve using matrix reference materials (AAS)*

The estimation of the contribution of the uncertainty in a reference material to the overall uncertainty in a calibration curve and, in turn, in the determined value of an element in a sample is not easy. An attempt by ISO/REMCO to resolve this problem has not been brought to a successful completion. Perhaps it is because of this difficulty why it is a widely applied practice in analytical laboratories that calibration curves are established with commercially-purchased or in-house prepared aqueous solutions of a simple, pure chemical salt and that matrix reference materials be that match as closely as possible the samples are measured with the samples to determine if there are any matrix or inter-element effects that must be resolved. In other words, the reference materials are used to assess the method (calibration curve) for accuracy.

6 SUMMARY

1. There are many important considerations and their impact on their use should be understood by a user in selecting appropriate matrix reference materials for use in assessing an analytical method for accuracy and precision.
2. The statistical tests in ISO Guide 33 "Uses of certified reference materials" to assess a method for accuracy and precision have been described and have been illustrated by application to two gold reference ores, MA-1b and CH-3.
3. The concept of longer term assessment of accuracy the repeated measurement of a given reference material over a period of time has been introduced to the user to provide a more demanding and sensitive statistical test where necessary.
4. The use of matrix reference materials in calibration was introduced and discussed in a general way. The use of the uncertainty of certified values in estimating the uncertainty of the calibration curve and, in turn, the determined value of an element in a sample is still unfolding.
5. Finally, the increasing pressure on reference material producers to have their quality system accredited to ISO Guide 34 "General requirements for the competence of reference materials producers" has not been touched on. ISO/REMCO and ILAC

(International Laboratory Accreditation Conference) are leading this initiative. However since accreditation to ISO Guide 34 is not available at this time nor in the near future, except in Australia and Japan, a user may want to select matrix reference materials from producers or have ISO 9000 certified and/or ISO Guide 25 accredited quality systems

References

ISO Guide 30 'Terms and definition use din connection with reference materials'
ISO Guide 31 'Contents of certificates, certification reports and labels of reference materials'
ISO Guide 32 'Calibration of chemical analyses and the use of certified reference Materials' - (Draft)
ISO Guide 33 'Uses of certified reference materials'
ISO Guide 34 'General requirements for the competence of reference materials producers'
CCRMP Catalogue of Certified Reference Materials, Report CCRMP 94-1E plus supplemental materials.

Acknowledgment

The author wishes to thank Clint Smith and Harry Alkema of Environment Canada for reviewing this manuscript and Diane Desroches for her assistance with the preparation of Figures 3 to 5. In addition, the author thanks the Analytical Services Group and the Canadian Certified Reference Materials Project for permission to use results and information from chemical analyses and reference materials.

The Selection and Use of Reference Materials: Some Examples Produced by LGC

LGC (TEDDINGTON) LTD, QUEENS ROAD, TEDDINGTON, MIDDLESEX, TW11 0LY, UK

1 THE ROLE OF REFERENCE MATERIALS IN CHEMICAL MEASUREMENTS

The global community is becoming increasingly dependent on reliable chemical measurements for sound decision making in a wide variety of areas, such as international trade, environmental protection, safe transportation, law enforcement, consumer safety and the preservation of human health. In these and other areas, poor quality data incurs a cost penalty, due not only to the need for repeat measurements, but also to the consequences of inappropriate actions based on false data. In the latter situation, the costs involved can be punitively high. The quality of all chemical measurements depends critically on the use of reliable reference materials for such purposes as the calibration of instrumentation, checking the performance of instrumentation against specification, the qualitative identification of the analyte or species being sought and the quantitative validation of measurement methodology being used.

It is because of the importance of reference materials in assuring the quality of chemical measurement data that the production of certified reference materials forms a major component of most countries chemical metrology programmes. Ultimately, the use of a fully documented certified reference material enables a laboratory to demonstrate the traceability of its results to a reliable measurement standard. If all measurement data, regardless of the measurement method used or the location of the laboratory carrying out the measurement, were traceable to an appropriate certified reference material, the mutual acceptance of data between different testing sites would be considerably enhanced.

Generally the demand for reference materials exceeds supply in terms of the range of materials and availability and it is rare to have a choice between two suppliers even though there are over 200 producers worldwide. As a result, the measurement scientist sometimes has to use the best available material but which may be a compromise to the ideal material required. It is, therefore, important that users and accreditation bodies understand the quality limitations of materials.

The absence of any internationally recognised classification, or approval system, can make life more difficult for the user. Not all materials that are used as reference materials are described as such. For example,commercially available chemicals of varying purity and products from research programmes are often used as standards or reference materials. In the absence of certification data provided by the supplier it is the responsibility of the user to assess the information available and undertake further characterisation as appropriate. Guidance on the preparation of reference materials is given in ISO Guides 31, 34 and 35 (Ref 1-3).

2 USES OF REFERENCE MATERIALS

Reference materials are used by the measurement scientist to investigate a wide range of applications in the laboratory including:

Method validation
Measurement uncertainty
Verification of the correct use of a method
Instrument calibration
Quality control
Traceability
Production of secondary RMs

Examples of some of these applications are given later in the text, but in all cases the user must assess the appropriateness and fitness for purpose of any CRM based on the customer and analytical requirements. Factors to be considered include:

Measurand (analyte)
Measurement range (concentration)
Matrix match and potential interferences
Measurement uncertainty
Certification procedures (measurement and statistical)
Quality management
Third party verification of technical and quality management issues
Completeness and transparency of information

In general, CRM's should be be used continuously in the frame-work of a comprehensive quality assurance system. It is not acceptable to use a CRM once only and then to assume that accurate results will be produced into the future. There are, however, cases for the legitimate use of CRM's on a non-routine basis e.g. where a new method has been developed by a laboratory and information is needed as to whether reliable and accurate results are being produced.

When CRM's are used, many users are unsure as to how best to interpret the results obtained for the reference material. This includes, how to compare the actual test results obtained for the CRM with the certified reference values and with their associated uncertainties quoted in the accompanying certificate.

The most common problems raised by analysts include:-

• How may replicate results are required for a proper comparison of the certified reference values and the actual analysis results?

• Is it necessary for the mean of the test results found for a CRM to lie within the uncertainty range of the certified value and are any differences found actually significant?

- If the mean of the test results found for a CRM lies outside the uncertainty range, what valid conclusions can be drawn and what action should be taken (if any)?

- Is a result outside the uncertainty range acceptable and if so what are the limits that do have to be respected. What are the conclusions and necessary actions in case of non-compliance with these limits?

All of these requirements may be specified in customer and analytical requirements, but often it will be necessary for the analyst to use professional judgement. The amount of evidence required to assess the suitability of a reference material depends on the requirement, but clearly a substantial amount of evidence will be required to demonstrate the suitability of a reference material for a high level and critical requirement.

Ideally, a certificate complying with ISO Guide 31 and a report covering the characterisation, certification and statistical procedures which comply with ISO Guide 35 will be available. However, many reference materials, particularly older materials and materials not specifically produced as reference materials, may not fully comply with ISO Guides 31 and 35. Alternative, equivalent information, in whatever form it is available, that provides credible evidence of compliance can be considered acceptable. Examples include the following: technical reports, trade specifications, papers in journals or reports of scientific meetings and correspondence with suppliers.

Assessment of quality requires the comparison of information concerning the appropriateness and fitness for purpose requirements with information about the candidate reference material available on a certificate, supporting report or from other sources. In the absence of specific information it is not possible to assess the quality of a reference material. The rigour with which the assessment needs to be conducted depends on the criticality and the level of the technical requirement and the expected influence of the specific reference material on the validity of the measurement. Only where the absence of a formal assessment can be expected to significantly affect measurement results is a formal quality assessment required and only key requirements must be documented. Laboratories must, however, be able to explain and justify the basis of selection of all reference materials.

3 BASIC CONSIDERATIONS WHEN USING A REFERENCE MATERIAL

In using a CRM to investigate a measurement application, the user will need to keep in mind the following basic rules in order to be able to interpret any data meaningfully:

(a) Independent replicate results are required

Independent replication requires that all the analytical stages of interest are carried out each time in order to obtain the individual results, *i.e.*, a replicate result is not influenced by previous replicate results. For example, independent replication could mean a completely new instrument calibration is required for each replicate determination in order to remove the effect of a common calibration.

(b) Data should not be rounded until after the statistical tests are carried out

Whenever repeated analytical measurements are made there is always some variation in the results caused by, for example, changes in environmental conditions or a slight variation in the way the sample is introduced into an instrument. If the data is rounded before carrying out any statistical test the natural variation within the data set can be masked (artificially increased or decreased). As a result the statistical tests may loose power or not work at all as demonstrated in Table 1.

Table 1

(i)	7.172, 7.175, 7.165, 7.183, 7.18	Mean = 7.175	Standard deviation = 0.007
(ii)	7.2, 7.2, 7.2, 7.2, 7.2	Mean = 7.2	Standard deviation = 0

(c) Always inspect the data before carrying out a statistical test

Extreme results or incorrect assumptions about how the data is distributed can lead to false conclusions being reached. It is, therefore, always a good idea to plot the data as a first stage in the analytical process.

Figure 1 *Blob plots of the raw data*

For small amounts of data (< 15 replicate measurements), a blob (or dot) plot can be used to explore how the data is distributed (Fig. 1). Blob plots are constructed simply by drawing a line, marking it off with a suitable scale and plotting the data along the axis. For larger data sets, frequency histograms (Fig. 2a) and Box and Whisker plots (Fig. 2b) may be better options to display the data distribution.

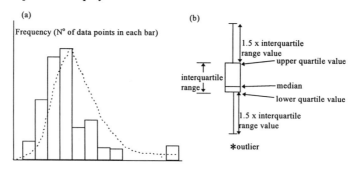

Figure 2 *Frequency histogram and Box and Whisker plot*

(d) The analytical determinations should be made using a method which is under statistical control

  Wrong conclusions about the accuracy and precision of a method can be reached if the method being checked/validated/calibrated is not under statistical control. The replicate results should therefore be plotted against a time index (*i.e.* the order the data was obtained, hours, days, etc.). If any systematic trends are observed (Figs. 3a to 3c) then the reasons for this must be investigated. Normal statistical methods assume a random distribution about the mean with time (Fig. 3d).

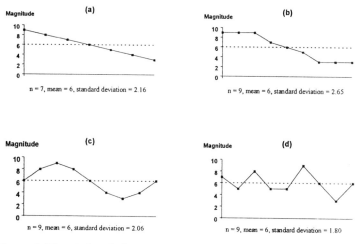

Figure 3 *Time indexed plots*

  To decide whether or not the mean of a set of replicate results is equivalent to the certified value of a CRM, the following four factors need to be considered:

1. The number of replicate determinations made on the CRM using the method under investigation.
2. The precision of the method under investigation.
3. The uncertainty in the certified value of the CRM.
4. The magnitude of the difference between the certified value of the CRM and the mean of the replicate results.

  Figure 4 show the uncertainty associated with the estimated standard deviation. From this plot it can be seen that to get useful information about the precision of any analytical method, between 6 and 20 independent replicates are required. The exact number within this range will depend to some extent on the precision of the method under investigation, *i.e.*, the more precise the method the fewer replicates are required.

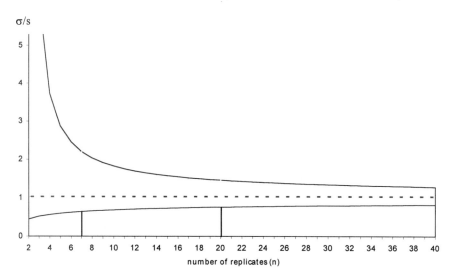

σ/s

number of replicates (n)

Figure 4 *Uncertainty associated with estimating a standard deviation*

It can be seen, therefore, that the sample mean and sample standard deviation calculated from the replicate measurements are only estimates of the real (population) mean and standard deviation. How good an estimate is dependent on the precision of the method and the number of replicate determinations made.

Detection of Outliers

Occasionally when carrying out repeated measurements one or two results may appear anomalous, *i.e.* extreme values. It is possible to check for the presence of these points of influence using one of the outlier tests described below (ideally also plotting the data).

Extreme values are defined as observations in a sample, so far separated in value from the remainder as to suggest that they may be from a different population, or the result of an error in measurement. Extreme values can also be subdivided into *stragglers* (extreme values detected between the 95% and 99% confidence level), and *outliers* (extreme values at >99% confidence level)

It is tempting to remove extreme values from a data set, because it is often believed that their retention will incorrectly alter the calculated statistics, *e.g.,* increase the estimate of precision, or possibly introduce a bias in the calculated mean. There is one golden rule, however, *no value should be removed from a data set on statistical grounds alone.* 'Statistical grounds' include outlier testing.

Outlier tests essentially flag up, on the basis of some simple assumptions, where there is most likely to be a technical error. They do *not* flag up that the point is 'wrong'. No matter how extreme a value is in a set of data, the suspect value could nonetheless be a correct piece of information. Only with experience or the identification of a particular cause can data be declared 'wrong' and removed. Most outlier tests look at some measure of the relative distance of a suspect point from the mean value. This measure is then assessed to see if the extreme value could reasonably be expected to have arisen by chance. Most of the tests look for single extreme values (Fig. 5a), but sometimes it is possible for several 'outliers' to be present in the same dataset. These can be identified in one of two ways:

- By iteratively applying the outlier test (not recommended when estimating the precision of an analytical method).
- By using tests which look for pairs of extreme values, *i.e.*, outliers that are masking each other (*see* Figs. 5b and 5c).

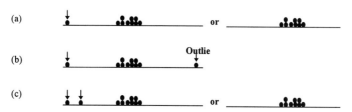

Figure 5 *Outliers and masking*

As a rule of thumb, if more than 20% of the data set is identified as outlying it becomes necessary to question basic assumptions about the data distribution and/or the quality of the data collected, *i.e.*, is the method is under statistical control.

The appropriate outlier tests for the three situations illustrated in Fig. 5 are: (a) Grubbs 1, (b) Grubbs 2, and (c) Grubbs 3. The test values for the three Grubbs' tests are calculated using equations 1 to 3, after arranging the data in ascending order.

$$G_1 = \frac{|\bar{x} - x_i|}{s} \qquad \text{(Eqn 1)}$$

$$G_2 = \frac{x_n - x_1}{s} \qquad \text{(Eqn 2)}$$

$$G_3 = 1 - \left(\frac{(n-3) \times s_{n-2}^2}{(n-1) \times s^2} \right) \qquad \text{(Eqn 3)}$$

where s is the standard deviation for the whole data set, x_i is the suspected single outlier *i.e.*, the value furthest away from the mean, $|\ |$ is the modulus - the value of a calculation ignoring the sign of the result, $\bar{x}$ is the mean, n is the number of data points, x_n and x_1 are the most extreme values and s_{n-2} is the standard deviation for the data set excluding the suspected pair of outlier values *i.e.*, the pair of values furthest away from the mean. If the test values (G_1, G_2, G_3) are greater than the critical value obtained from standard statistical tables, then the extreme value(s) are unlikely to have occurred by chance at the stated confidence level.

As noted above, no value should be discarded based only on statistical information. For example, Fig. 6 shows three situations where outlier tests can misleadingly identify an extreme value.

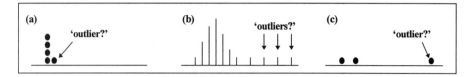

Figure 6 *Pitfalls of outlier testing*

Fig. 6a shows a situation common in chemical analysis. Because of limited measurement precision (rounding errors) it is possible to end up comparing a result which, no matter how close it is to the other values, is an infinite number of standard deviations away from the mean of the remaining results. This value will therefore always be flagged as an outlier.

In Fig. 6b there is a genuine long tail on the distribution which may cause successive outlying points to be identified. This type of distribution is surprisingly common in some types of chemical analysis, particularly trace analysis, e.g. pesticide residues in environmental matrices.

If there are few measurement results (Fig. 6c), an outlier can be identified by chance. In this situation it is possible that the identified point is closer to the 'true value' and it is the other values that are the outliers. This occurs more often than we would like to admit.

Worked Example

13 independent replicate determinations of the nickel content (mg/kg) were carried out on a soil contaminated with clinker/ash reference material (LGC CRM6141). The results are arranged in ascending order:

47.876, 47.997, 48.065, 48.118, 48.151, 48.211, 48.251, 48.559, 48.634, 48.711, 49.005, 49.166, 49.484

$$n = 13, \text{ mean} = 48.479, s = 0.498, s_{n-2}^2 = 0.123$$

$$G_1 = \frac{49.484 - 48.479}{0.498} = 2.02 \qquad G_2 = \frac{49.484 - 47.876}{0.498} = 3.23$$

$$G_3 = 1 - \left(\frac{10 \times 0.123}{12 \times 0.498^2} \right) = 0.587$$

Grubbs' critical values for 13 values are $G_1 = 2.331$ and 2.607, $G_2 = 4.00$ and 4.24 and $G_3 = 0.6705$ and 0.7667 for the 95% and 99% confidence levels respectively. Since, in all cases, the test values are less than their respective critical values it can be concluded there are no outlying values.

4 USE OF REFERENCE MATERIALS FOR ASSESSING THE ACCURACY OF TEST DATA

Even when a laboratory is using an analytical method which has documented performance characteristics for trueness and precision that have been established according to recognised method validation protocols, there is still a need for a practical demonstration by the laboratory that the accuracy of its results on routine samples are acceptable and fit-for-purpose. The need is even greater if the analytical method being used has not been the subject of a full validation exercise. There are various approaches to demonstrating that routine results are of the required accuracy (e.g. participation in proficiency testing schemes, the analysis of 'spiked' samples, using two or more methods and sample exchanges with other laboratories), but the analysis of a suitable matrix reference material offers a particularly effective means of assessing and demonstrating the quality of routine analytical data. By comparing the result obtained uisng the analytical method with the RMs value, any bias in the result can be detected and, if necessary, appropriate corrective action can be taken. It is important to note that when used in this way matrix reference materials enable any bias due to a laboratory's execution of a particular method to be identified. Of course, there is an implicit assumption that the method itself is capable of producing unbiased data.

When using a matrix reference material for this purpose, the matrix of the reference material should resemble as closely as possible the matrix of the routine samples being analysed. In this way, the problems posed by the routine samples due to such matters as analyte speciation, analyte extraction efficiency, completeness of matrix dissolution and interferences in analyte quantitation from matrix components are replicated by the reference material. Under these circumstances a result for the reference material that adequately agrees with the documented value gives good grounds for concluding that the results on the routine samples are fit-for-purpose.

Although the matrix of the reference material should, ideally, be identical to that of the routine samples, this will rarely be possible. For example foodstuff reference materials may have been produced in the freeze-dried form (to enhance their stability) which is markedly different from the form of the fresh foodstuffs that may be undergoing analysis. Likewise, environmental reference materials such as soils, sediments etc. may have been dried to a particular moisture content and ground to a specific particle size which could be different from that which a laboratory would normally use in the preparation of its routine samples. In yet other types of matrix materials such as drinking waters, blood serum etc, chemical stabilisers may have been added by the producer of the material, which would not normally be present in the routine samples. Users should therefore be aware that differences of the general type outlined above may affect the analytical procedure to the extent that the results obtained on the reference material are not a good guide to the accuracy of the routine results. It is always the responsibility of the user to ensure that the matrix reference material selected is appropriate for its intended use.

Essentially, the accuracy of an analytical result is characterised by the closeness of the agreement between the result and the true value. The true value is the actual concentration of an analyte in a sample matrix and it is the value that would be obtained by a perfect measurement. True values, by their nature, are indeterminate but the values of reference materials (and particularly certified reference materials) are considered to be valid estimates of the true values, within the stated uncertainties.

Broadly speaking, two factors may be regarded as contributing to the accuracy of a result:

- the precision of the analytical method, as executed (random error);
- the trueness (i.e. lack of bias) of the analytical method, as executed (systematic error).

The prime use of a matrix reference material is to assess a result for the presence of bias (systematic error). The bias (Δ) of an analytical result ($\bar{x}$) is given by:

$$\Delta = \bar{x} - \mu \qquad\qquad \text{(Eqn 4)}$$

where μ = true value.

If bias is assessed by analysis of a matrix reference material, μ = the documented property value of the reference material.

Because of the presence of random error, the value of Δ is likely to be some value other than 0, even when there is no bias present in the result $\bar{x}$. The larger the random error, the larger the value of Δ can be without it being significantly different from 0. That is, the larger the random error, the more difficult it is to detect relatively low levels of bias.

4.1 Assessment of Precision

The precision of a measurement result may be regarded as arising from two contributing factors:

- the within-laboratory repeatability standard deviation of the method (s_w);
- the between-laboratory reproducibility standard deviation of the method (s_b).

Both components of precision should be taken into account when the bias of an analytical result is assessed. The within-laboratory repeatability standard deviation (s_w) may be estimated from the replicate measurements made on the reference material as part of the bias assessment. To obtain a reliable estimate of s_w, it is recommended that at least 6 replicate measurements are carried out under repeatability conditions. The term 'repeatability conditions' means that the measurements are carried out by the same analyst, using the same method and equipment, in the same laboratory environment over the shortest practical timescale. The between-laboratory component of the precision is less easy to estimate. The following approaches are suggested, in order of preference:

1. The long-term standard deviation (intermediate precision) of a set of replicate measurements (at least 6 and preferably up to 20) made on the reference material (or similar matrix) over an extended period of time (say >3 months) in one laboratory often provides an acceptable approximation to the between-laboratory reproducibility standard deviation (s_b).

2. If the analytical method being used has been the subject of an interlaboratory validation exercise conducted according to recognised protocols, the values for the repeatability standard deviation (s_r) and reproducibility standard deviation (s_R) quoted in the standard document for the method may be used to estimate s_b, thus:

$$s_b = \sqrt{(s_r^2 + s_R^2)}$$

3. Where the reference material has been characterised by means of an interlaboratory exercise, relevant information on the between-laboratory standard deviation may be given on the documentation accompanying the material. If the analytical method being used by the laboratory wishing to assess its individual bias is similar to the method(s) used in the characterisation of the reference material, any between-laboratory standard deviation quoted in the reference material documentation could be adopted as an estimate of s_b.

4. The interlaboratory standard deviation predicted by the Horwitz function provides a further means for estimating s_b.

5. In the absence of any other information, an estimate of s_b may be obtained from the measured s_w value, thus:

$$s_b \approx 2 \text{ to } 3 \times s_w$$

The precision (σ) of a laboratory's measured result on a matrix reference material is calculated by combining the two components in the following manner:

$$\sigma = \sqrt{s_b^2 + \frac{s_w^2}{n}} \qquad \text{(Eqn 5)}$$

n = number of replicate measurements made on the reference material.

Generally speaking, s_w may be expected to be smaller than s_b (typically by a factor 2 to 3). Combined with the fact that n will normally be at least 6, this means that s_b is usually the major contributor to σ, to the extent that s_w can usually be ignored.

At first sight it mat seem that the estimated precision of an *individual* laboratory's result, σ, should include a contribution from the between-laboratory variation, s_b. Should not the within-laboratory variation, s_w, actually measured by replicate determinations on the matrix reference material, be an adequate measure of the expected random dispersion of measured results about the true value?

However, it should be appreciated that s_w measures the random dispersion of a laboratory's replicate results about the mean of those results. The mean itself is randomly distributed about the true value for the reference material with a dispersion that is characterised by s_b. Thus, the value of s_w and s_b combined (as in equation 5) is used to describe the overall dispersion of results about the true value.

The parameter s_b measures those sources of random error that cannot be evaluated by replicate measurements in a single laboratory, but which still contribute to the dispersion of an individual laboratory's result about the true value. An example of such a source of random error would be the final volume of a prepared sample extract, before it is introduced to an instrument for measurement, where no account was taken of the ambient temperature in the laboratory (a common occurrence). For measurements made under repeatability conditions, variations in ambient temperature would be insignificant and therefore not included in s_w. However, the same measurements made in different laboratories (or indeed in a single laboratory over a period of time) could be the subject of additional random error due variations in ambient temperature. The effect of such variations would be included in s_b.

It may also be helpful to realise that when a laboratory analyses a matrix reference material it is effectively taking part in an "interlaboratory comparison", as the documented value for the reference material will usually have been established by measurements carried out in one or more other laboratories. Under these circumstances it is obviously appropriate that between-laboratory component of precision should be taken into account when an individual laboratory compares its result to the documented value for the reference material.

If information is available on the within-laboratory standard deviation expected for the method (e.g. the repeatability standard deviation (s_r) quoted in the standard document for a method validated in an interlaboratory exercise according to recognised protocols), then the Chi-squared test may be carried out:

$$\chi_c^2 = \left(\frac{s}{\sigma}\right)^2 \qquad\qquad \text{(Eqn 6)}$$

where σ is the required precision value for the method expressed as a standard deviation, s is the determined method precision expressed as a standard deviation and χ_c^2 is the calculated Chi-squared test value. The critical Chi-squared value is calculated as follows:

$$\chi_{crit}^2 = \frac{\chi_{(n-1);0.95}^2}{n-1} \qquad\qquad \text{(Eqn 7)}$$

where the value of $\chi_{(n-1);0.95}^2$ is found from statistical tables.

If $\chi_c^2 \leq \chi_{crit}^2$ there is no evidence that the precision of analytical method is inadequate.

If $\chi_c^2 > \chi_{crit}^2$ there is evidence that the precision of analytical method is inadequate.

Worked Example

In the potentiometric titration of an 80mg/100mL forensic alcohol CRM (LGC CRM5401), the required precision, expressed as a standard deviation is 0.1 mg/100mL. The following results were obtained from 16 replicate determinations:

79.99, 80.06, 80.04, 79.96, 80.03, 80.03, 80.01, 79.90, 80.03,
80.01, 79.89, 80.00, 79.89, 79.94, 79.98, 79.95

Mean = 79.98124 mg/100mL, Standard deviation = 0.05450 mg/100mL.

$$\chi_c^2 = \left(\frac{0.05450}{0.1}\right)^2 = 0.2970 \qquad\qquad \chi_{crit}^2 = \frac{24.999}{15} = 1.6666$$

Since $\chi_c^2 \leq \chi_{crit}^2$, there is no evidence that the precision of the potentiometric titration method for ethanol is inadequate.

If the precision of the method is found to be inadequate then the method should be looked at in detail to identify those areas were most variation is occurring. By controlling these areas first the greatest improvement in the methods precision can be achieved.

4.2 Assessment of Bias

Bias (systematic error) can be assessed using significance testing by comparing the certified value, μ, with the mean result from replicate determination of the CRM, $\bar{x}$, analysed using the method under investigation (a_1 and a_2 may, if considered appropriate, be included in the equation to take account of any technical considerations such as a known bias in the method):

$$-a_2 - 2\sigma - \leq \bar{x} - \mu \leq a_1 + 2\sigma \qquad \text{(Eqn 8)}$$

Equation 8 assumes the uncertainty in the certified value is insignificant compared with the method precision. If this is not the case then the bias assessment criteria become:

$$-a_2 - 2\sqrt{u^2_{CRM} + \sigma^2} - \leq \bar{x} - \mu \leq a_1 + 2\sqrt{u^2_{CRM} + \sigma^2} \qquad \text{(Eqn 9)}$$

where u_{CRM} is the standard uncertainty for the reference material.

As a result of the formula used to detect bias (equation 9), bias detection is limited to a minimum of twice the standard uncertainty of the certified value of the CRM. It is therefore incumbent on the user to select a reference material which is capable of detecting a possible significant (i.e. unlikely to have occurred by chance) bias in the method being assessed.

Like most statistical tests, it is assumed that the sample correctly represents the population and that the population follows a normal distribution. Although these assumptions are never exactly complied with, in a large number of situations where laboratory data is being used they are not grossly violated.

Worked Example

The long-term precision of a method used to determine lead in drinking water using a CRM check sample was estimated as 1.17 µg/kg with a mean of 57.6 µg/kg. The certified value for the CRM is 60.1 µg/kg with an expanded uncertainty of 2.5 µg/kg. A coverage factor of k=2 has been used to calculate the expanded uncertainty. Is there any evidence of a bias in the determination?

$$\bar{x} - \mu = 57.6 - 60.1 = -2.5 \qquad \text{and} \qquad 2\sqrt{u^2_{CRM} + \sigma^2} = 2 \times \sqrt{\left(\frac{2.5}{2}\right)^2 + 1.17^2} = 3.42$$

Since $-3.42 \leq -2.5 \leq +3.42$, there is no evidence of a significant bias in the method.

Note:

1. The expanded uncertainty quoted on the certificate needs to be converted into a standard uncertainty (standard deviation). This is achieved by dividing the expanded uncertainty by the coverage factor *i.e.*, $\frac{2.5}{2}$.

2. It is assumed that enough replicates have been determined to obtained a good estimate of the methods precision, σ^2.

If a discrepancy is found between the user's measurement and the certified value this needs to be investigated. In most cases the method can be adjusted to remove an unsuspected bias. The user should also consider the possibility that the CRM has been stored incorrectly and/or has become contaminated in the laboratory. In the exceptional circumstance were no reason can be found for the observed bias, the user should contact the producer of the CRM who should be able to advise the user further.

Worked Example

The reference material LGC1004 comprises a water matrix containing triazine herbicides at certified concentrations. For the herbicide simazine, the certified concentration is 26.7µg/kg, with an expanded uncertainty (k=2) of 2.0µg/kg. Six replicate analyses were carried out on this material, with the following results:

$$29.4, 24.9, 26.4, 25.7, 22.0, 23.5$$
$$n = 6, \text{mean} = 25.3, s_w = 2.5$$

The value adopted for s_b is 5.2µg/kg, based on the measured long term repeatability of the procedure. The calculated value for σ is given by:

$$\sigma = \sqrt{5.2^2 + \frac{2.5^2}{6}} \qquad = \quad 5.3\mu g/kg$$

The calculated value for the apparent bias is given by 27.3 - 26.7 = + 0.6 µg/kg. It is seen that the measured value for the apparent bias meets the condition of equation 8:

$$-10.6 \ < \ 0.6 \ < \ 10.6$$

It may therefore be concluded that there is no evidence for the presence of bias in the laboratory's execution of the method. It should be noted that this is **not** the same as concluding that the method and its execution is unbiased. There could be bias present that is beyond the power of the test to detect. It should also be noted that the validity of the test depends entirely on the validity of the values adopted for s_b and s_w. If these values are wrong the value for σ will also be wrong and the test will lead to false conclusions being drawn.

4.3 Should the result obtained on a CRM lie within its uncertainty range?

It is sometimes concluded that if a result obtained by a laboratory on a matrix reference material falls outside the uncertainty range of the documented property value, then the result must be biased. However, this conclusion is false as it takes no account of the random uncertainty (σ) in the result as obtained by the laboratory. The uncertainty range of the reference material only addresses the uncertainties arising in the characterisation of the reference material; it does **not** deal with the uncertainty associated with measurements made on the reference material in the user's laboratory.

Worked Example

The reference material LGC7103 comprises a sweet digestive biscuit matrix with certified values for the concentrations of various nutritional components. The certified value for the nitrogen (protein) content is 1.08g/100g, with an expanded uncertainty (k=2) of 0.01g/100g. Six replicate measurements were carried out on the material with the following results:

$$1.091, 1.094, 1.097, 1.099, 1.100, 1.095$$
$$n = 6, \text{ mean} = 1.096, s_w = 0.004$$

The value adopted for s_b is 0.015 g/100g, based on the long term repeatability of the method. The calculated value for σ is given by:

$$\sigma = \sqrt{0.015^2 + \frac{0.004^2}{6}} \qquad = 0.015 \text{ g/100g}$$

The calculated value of the apparent bias is $1.096 - 1.08 = 0.016$ g/100g. It is seen that the apparent bias meets the requirements of equation 8, viz:

$$-0.03 < 0.016 < 0.03$$

It may be concluded, therefore, that there is no evidence for the presence of bias in the method, as executed by the laboratory. However, it should be noted that the measured value of 1.096 lies **outside** the uncertainty range for the reference material, which is 1.08 ± 0.01, confirming that it is **not** a requirement that all results must fall within this range if it is to be concluded that there is no evidence for saying a method is biased.

Although the uncertainty range of a reference material's certified value(s) does not necessarily define the acceptable limits for results obtained by users of the material, the uncertainty of the reference value is used in other ways. In particular, it is important to establish the significance of the uncertainty of the documented reference value in relation to the estimated uncertainty (σ) of result obtained on the reference material by the user laboratory. In those situations where the uncertainty of the reference value is likely to be significant in relation to the estimated uncertainty (σ) of the result obtained on the material by a user laboratory, the uncertainty of the reference material must be taken into account when the result is assessed for the presence of bias.

Rather than simply using the value σ (as defined by equation 5) to calculate the acceptable range of results (from equation 8), an additional term needs to be included in the calculation of σ. The additional term represents the standard uncertainty of the documented value of the reference material. The standard uncertainty of the reference material is calculated from the documented uncertainty, which is normally an expanded uncertainty (U) with a stated coverage factor (k). If the coverage factor used is 2, the true value is thought to lie in the range given by the expression (documented value $\pm$ expanded uncertainty) with a 95% level of confidence. If the coverage factor used is 3, the level of confidence is 99% The standard uncertainty (u) of the reference material is calculated as:

$$u_{RM} = U/k$$

The value of σ used to calculate the acceptable range of results is then given by:

$$\sigma = \sqrt{s_b{}^2 + \frac{s_w{}^2}{n} + u_{RM}{}^2} \qquad\qquad \text{(Eqn 10)}$$

It follows from equation 10 that if $u_{RM}{}^2$ is equal to or less than $0.1 \left(s_b{}^2 + \frac{s_w{}^2}{n} \right)$, then the

uncertainty of the reference material makes a negligible contribution to the calculation of σ and may be ignored. This condition will be fulfilled when u_{RM} is equal to or less than one third of the estimated standard uncertainty $\sqrt{s_b{}^2 + \dfrac{s_w{}^2}{n}}$ of the laboratory's measured result for the reference material.

Worked Example

The reference material LGC7131 is a soft drink with certified levels of artificial sweeteners. The certified concentration for saccharin is 76mg/litre with an expanded uncertainty (k=2) of 7 mg/litre. The results of 6 replicate measurements on the material are

$$81.46, 81.64, 80.47, 81.59, 81.25, 80.99$$
$$n = 6, \text{ mean} = 81.23, s_w = 0.44$$

The value adopted for s_b is 1.5 mg/litre, based on the long term repeatability of the method. The value for σ (based on eq 5.2) is given by: $\sigma = \sqrt{1.5^2 + \dfrac{0.44^2}{6}} = 1.51$. The calculated value for the apparent bias (Δ $= \overline{X} - \mu$) is $81.23 - 76 = +5.23$ mg/litre.

It can be seen that this value does **not** meet the requirements of equation 8 in that $\overline{X} - \mu$ (5.23 mg/litre) is **greater** than 2σ (3.02 mg/litre). On these figures, therefore, it must be concluded, at the 95% confidence level, that there **is** evidence for the presence of bias in the method. However, comparison of the estimated value for σ (1.51), with the standard uncertainty for the certified concentration value shows that the latter cannot be ignored:

$$u_{RM} = 7 \div 2 = 3.5 \text{ mg/litre.}$$

Rather than u_{RM} being less than one third of the value of σ estimated as above, it is actually greater than σ. We therefore need to calculate σ using equation 10, rather than equation 5. The appropriate value for σ is therefore given by:

$$\sigma = \sqrt{1.5^2 + \frac{0.44^2}{6} + 3.5^2} = 3.81 \text{ mg/litre}$$

It is now seen that the observed value for the apparent bias of 5.23 mg/litre **does** comply with the requirements of equation 5.3, in that 5.23 is now less than the 2σ value ($2 \times 3.81 = 7.62$ mg/litre). It may therefore be concluded that, taking account of the uncertainty of the documented value of the reference material, there is no evidence for the presence of bias in the method, as executed by the laboratory.

5 USE OF REFERENCE MATERIALS TO DETECT OTHER FORMS OF BIAS

In addition to using a significance test to detect bias in a single method using a CRM, such tests can also be used to see if there is a difference between:

- the mean of replicate results and a regulatory limit.
- two individual analysts.
- two different methods.
- two laboratories within a single organisation.
- two laboratories in different organisations.
- the spread/dispersion of two set of replicate data.

Although CRMs are not essential to answer the types of question posed above, in all but the first and last case it is necessary to use standards with well characterised matrices and a known or controlled level of homogeneity. Of course, CRMs and RMs meet these requirements.

t-Tests

When comparing the mean result of a number of replicate determinations with another mean or regulatory limit, t-tests are used. There are a number of these, the formulae for each are given in Table 2. To carry out a *t*-test the following steps are required:

Table 2 *t-Test Formulae*

t-test to use when comparing	Equation
The long term average or regulatory limit ($\bar{x}_0$) with a sample mean	$$t = \frac{\bar{x} - \bar{x}_0}{s/\sqrt{n}}$$ The degrees of freedom v is given by: $v = n_1 - 1$
The difference between pairs of analytical results (e.g. results obtained from two analytical methods)	For a two-tailed test, $t = \dfrac{\lvert\bar{d}\rvert \times \sqrt{n}}{s_d}$ The degrees of freedom v is given by: $$v = n_1 - 1$$
Difference between independent sample means(assuming equal variance)	$$t = \frac{\bar{x}_1 - \bar{x}_2}{s_c\sqrt{\left(\dfrac{1}{n_1} + \dfrac{1}{n_1}\right)}}$$ The degrees of freedom v is given by: $$v = n_1 + n_2 - 2$$ and $$s_c = \sqrt{\frac{s_1^2(n_1 - 1) + s_2^2(n_2 - 1)}{(n_1 + n_2 - 2)}}$$
Difference between independent sample means (assuming unequal variance)	$$t = \frac{\bar{x}_1 - \bar{x}_2}{\sqrt{\dfrac{s_1^2}{n_1} + \dfrac{s_2^2}{n_2}}}$$ The degrees of freedom v is given by: $$\frac{1}{v} = \frac{s_1^4}{k^2 n_1^2(n_1 - 1)} + \frac{s_2^4}{k^2 n_2^2(n_2 - 1)}$$ where $$k = \frac{s_1^2}{n_1} + \frac{s_2^2}{n_2}$$

$\bar{x}$ is the sample mean, μ is the population mean, s is the standard deviation for the sample, n is the number items in the sample, $\lvert\bar{d}\rvert$ is the absolute difference between paired means, $\bar{d}$ is the difference between paired means, s_d is the sample standard deviation for the pairs, $\bar{x}_1$ and $\bar{x}_2$ are two independent sample means, n_1 and n_2 are the number of items making up each sample, s_c is the combined standard deviation and s_1 and s_2 are the sample standard deviations for the two independent sample means.

Note: *For a one tailed t-test the sign of the calculated t value is important, because it shows the direction of the difference. For a two-tailed t-test the sign of the calculated t value is ignored.*

(a) State exactly what question is being asked. This is done in the form of two hypothesis; (i) the results being compared could really be the same, (ii) the results could really be different. This is written as:

The null hypothesis (H_0): $\mu_1 = \mu_2$;

The alternative hypothesis (H_1): $\mu_1 \neq \mu_2$. (two-tailed) or $\mu_1 < \mu_2$. (one-tailed) or $\mu_1 > \mu_2$. (one-tailed).

(b) Decide which of the *t*-tests listed in Table 2 is appropriate and calculate the *t*-value/statistic.

(c) Look up the critical *t*-value in standard t-test statistical tables. To do this the following information is required:

(i) Is the direction of any difference between the two values being compared critical (one-sided) or is it only important to check for a statistically significant difference (two-sided)?

(ii) The degrees of freedom for the appropriate *t*-test, calculated using Table 2.

(iii) How certain do the conclusions need to be? It is normal practice in chemistry to select the 95% confidence level (although in some situations this is an unacceptable level of error, *e.g.* in medical research, where the 99% or even the 99.9% confidence level can be chosen).

Using this information the critical *t*-value for the test can then be found from standard t-test statistical tables. For example, the critical *t*-value for a two-tailed test carried out at the 95% confidence interval with 14 degrees of freedom (15 replicate determinations) is 2.145.

Worked Example

A chemist is asked to determine if the results from a sample exceeds the regulatory limit of 22.7 mg/dm^3 (i.e. in this example it is necessary to test against a set limit). The mean of 10 replicate results is 23.5 mg/dm^3, with a standard deviation of 0.9 mg/dm^3. (It is assumed that the analyst is competent and reports unbiased results, this should be tested using a CRM as a check standard).

To answer this question we use the *t*-test to compare the mean value with a set limit.

The null hypothesis (H_0): $\overline{x}_0 = $ set limit.

The alternative hypothesis (H_1): $\overline{x}_0 > $ set limit.

$$t\text{ - value} = \frac{23.5 - 22.7}{0.9 / \sqrt{10}} = +2.81$$

Because we are only interested in whether or not the sample is above the regulatory limit we carry out a one-tailed *t*-test.

$t_{crit} = 1.83$ one-tailed at the 95% confidence level for 9 degrees of freedom.

Since $t_{calculated} > t_{crit}$ we can reject the null hypothesis and conclude that the sample is significantly above the regulatory limit.

The F-test

The F-test, compares the spread of results in two data sets. This is done in order to determine if they could reasonably be considered to come from the same parent distribution. The test can therefore be used to answer questions like:

- are the results from two different analysts comparable?
- are two methods equally precise?
- can we combine two estimates of spread (dispersion)?

The measure of spread used in the F-test is variance which is simply the square of the standard deviation. The variances are ratioed to get the test value:

$$F = \frac{s_a^2}{s_b^2} \qquad \text{(Eqn 11)}$$

This value is then compared with a critical value in standard F-test statistical tables which tells us how big the ratio need to be to rule out the difference in spread occurring by chance. The F_{crit} value is found from tables using (n_1-1) and (n_2-1) degrees of freedom.

Note:
(1) If we are interested in the question; is there a significant difference in the spread of the two sets of replicates (*i.e.*, a two-tailed F-test)? Then it is normal practice to arrange s_1 and s_2 so that F is > 1 and look up the 97.5% confidence level F critical value.
(2) If we are interested in the question; is the new method the same as or significantly more precise than the old method (*i.e.*, a one-tailed F-test)?
Then;

$$F = \frac{s_{new\,method}^2}{s_{old\,method}^2} \qquad \text{(Eqn 12)}$$

In this case the two precision estimates are not rearranged and the 95% confidence level F critical value is used. In either case, the standard deviations are considered to be significantly different if $F > F_{crit}$.

Worked Example

Two independent methods for determining the concentration of selenium in soil are to be compared with each other using a natural matrix environmental CRM . The results from each method are shown in Table 3:

Table 3

						$\bar{x}$	s
Method 1	4.2	4.5	6.8	7.2	4.3	5.40	1.471
Method 2	9.2	4.0	1.9	5.2	3.5	4.76	2.750

Using the *t*-test for independent sample means, we define the null hypothesis H_0 as $\mu_1 = \mu_2$, *i.e.* there is no difference between the means of the two methods (the alternative hypothesis is H_1: $\mu_1 \neq \mu_2$). If the two methods have standard deviations which are not significantly different (checked using the F-Test), then we can combine (or pool) the standard deviation (S_c) and use the t-test for equal variance:

$$F = 2.75^2 / 1.471^2 = 3.49 \qquad\qquad F_{crit} = 9.605 \text{ (5-1) and (5-1) degrees of freedom}$$

Since $F_{crit} > F_{calculated}$, we can conclude that the spread of results in the two data sets are not significantly different and it is therefore reasonable to combine the two standard deviations:

$$S_c = \sqrt{\left(\frac{1.471^2 \times (5-1) + 2.750^2 \times (5-1)}{(5+5-2)}\right)} = 2.205$$

Evaluating the test statistic t:

$$t \approx \frac{(5.40 - 4.76)}{2.205\sqrt{\left(\frac{1}{5} + \frac{1}{5}\right)}} \quad\Rightarrow\quad t \approx \frac{0.64}{2.205 \times 0.632} \approx \frac{0.64}{1.395} \approx 0.459$$

The 95% two-sided critical value is 2.306 for $v = 8$ $(n_1 + n_2 - 2)$ degrees of freedom. This exceeds the calculated value of 0.459, thus the null hypothesis (H_0) cannot be rejected and we conclude there is no significant difference between the results given by the two methods, for the standard being tested.

Worked Example

Two methods are available for determining the concentration of a vitamin in a foodstuff. In order to compare the methods (paired difference), several CRM with different matrices are prepared for analysis using the same extraction technique. Each sample preparation is then divided into two aliquots and readings are obtained on them using the two methods, ideally commencing at the same time in order to lessen the possible effects of sample deterioration. The results are shown in Table 4:

Table 4

Sample/Method	1	2	3	4	5	6	7	8
A	2.52	3.13	4.33	2.25	2.79	3.04	2.19	2.16
B	3.17	5.00	4.03	2.38	3.68	2.94	2.83	2.18
Difference (d)	-0.65	-1.87	0.30	-0.13	-0.89	0.10	-0.64	-0.02

The null hypothesis is $H_0 : \varsigma = 0$ against the alternative $H_1 : \varsigma \neq 0$ Where ς is   the population paired difference.

The test is a two tailed test as we are interested in both $\bar{d} < 0$ and $\bar{d} > 0$.

The mean, $\bar{d}$, = -0.475, and the sample standard deviation for the paired differences, $s_d = 0.700$.

$$t = \frac{|0.475| \times \sqrt{8}}{0.700} = 1.918$$

The tabulated value of t_{crit} (with $\upsilon=7$, at the 95% two-sided confidence limit) is 2.365. Since the calculated value is less than the critical value, H_0 cannot be rejected and it follows that there is no statistically significant difference between the two techniques.

6   USING A REFERENCE MATERIAL TO ESTIMATE A METHOD'S RECOVERY EFFICIENCY

A certified reference material with a matrix and analyte concentration representative of those which will routinely be analysed may be used to estimate the mean recovery of a method. At least 10 portions of the reference material are analysed in a single batch (if it is impractical to carry out 10 analyses in a single batch, the replicates should be analysed in the minimum number of batches possible over a short period of time) and each portion must be taken through the entire analytical procedure. The mean recovery, $\overline{R}_m$, is calculated as follows:

$$\overline{R}_m = \frac{\overline{C}_{obs}}{C_{CRM}}$$

where $\overline{C}_{obs}$ is the mean of the results from the replicate analysis of the CRM and C_{CRM} is the certified value for the CRM. The uncertainty in the recovery, $u(\overline{R}_m)$, is then calculated using:

$$u(\overline{R}_m) = \overline{R}_m \times \sqrt{\left(\frac{s_{obs}^2}{n \times \overline{C}_{obs}^2}\right) + \left(\frac{u(C_{CRM})}{C_{CRM}}\right)^2}$$

where s_{obs} is the standard deviation of the results from the replicate analyses of the CRM, n is the number of replicates and $u(C_{CRM})$ is the standard uncertainty in the certified value for the CRM.

The above calculation provides an estimate of the mean recovery of a method and its uncertainty. The contribution of recovery and its uncertainty to the combined uncertainty for the method depends on whether the recovery is significantly different from 1 (100%), and if so, whether or not a correction is made.

Calculating standard uncertainties

It should be borne in mind when estimating uncertainty that the information available (e.g. calibration certificates) may not be expressed in the form of a *standard* deviation and it must therefore be converted before it can be combined with other standard uncertainties. Some common cases are given below.

(a) The uncertainty is expressed as a confidence interval with a given level of confidence. For example, a CRM certificate states that the concentration of an analyte in a certified reference material is 100 ± 0.5 mg.kg^{-1} with 95% confidence.

To convert to a standard uncertainty, divide by the appropriate percentage point of the normal distribution for the level of confidence given. For 95% confidence divide by 1.96.

(b) The uncertainty is expressed as an expanded uncertainty calculated using a given coverage factor. For example, documentation supplied with a solution states that its concentration is 1000±3 mg.L , where the reported uncertainty is an expanded uncertainty calculated using a coverage factor $k = 2$ which gives a level of confidence of approximately 95%.

To convert to a standard uncertainty, divide by the stated coverage factor.

(c) Limits of ±x are given without a confidence level or coverage factor. For example, the manufacturer's specification for the stated volume of a 100 mL volumetric flask is quoted as ±0.08 mL. It is normally appropriate to assume a rectangular distribution with a standard deviation of $x/\sqrt{3}$.

To convert to a standard uncertainty, divide by $\sqrt{3}$.

7 USING A REFERENCE MATERIAL FOR INSTRUMENT CALIBRATION

Modern instrumental methods of analysis offer a wide range of potential benefits such as low detection limits, high specificity, good precision and automated sample throughput. However, in most cases the relationship between the output signal of an instrument (e.g. peak area, counts, mV, etc.) and the quantity of analyte being measured (e.g. g, mole, etc. is empirical in nature. There is no well-understood physical or chemical theory that describes the magnitude of the signal in terms of the quantity of analyte present. Consequently, the amount of analyte present in a test sample cannot be determined from the measured instrument signal from first principles.The utility of most analytical instruments arises entirely from the experimental observation that the observed instrument signal is some arithmetical function of the quantity of analyte. This function may often be described by:

$$\text{Signal} = \text{K x (analyte quantity)}^n \qquad \text{(Eqn 13)}$$

For the commonly encountered linear relationship between signal and analyte quantity, the value of n =1. The proportionality constant K is, of course, usually unknown, due to the lack of an appropriate physical/chemical theory to underpin the fundamental operation of the instrument.

Under these circumstances, it is necessary to calibrate the output signal of the instrument by introducing to the instrument accurately known quantities of the particular analyte(s) of interest. The signal thus obtained from the calibrant is then compared with the signal obtained from the test sample and the quantity of analyte in the sample is often determined by a calculation of the type:

Analyte quantity in test sample = $\dfrac{\text{Sample signal}}{\text{Calibrant signal}}$ x   amount of analyte in calibrant   (Eqn 14)

This calculation would apply where the instrument signal varied linearly with analyte quantity (i.e. n = 1 in Eqn. 13). It will be readily apparent that the validity of the calculation represented by Eqn. 14 depends on the values for n and K in Eqn. 13 being the same for both the calibrant and the test sample. In other words, the analyte in the calibrant must respond quantitatively to the instrument in the same manner as the analyte in the test sample. Only then will it be possible to compare like with like. In any other circumstance the comparison of the calibrant signal with the test sample signal will not be valid and an erroneous analytical result will be produced. Therefore it is necessary to establish that the experimental conditions used to calibrate an instrument are appropriate for the test sample that is to be ultimately analysed. The proper selection and use of appropriate reference materials as instrument calibration standards (solutions) is therefore extremely important and is discussed in detail below.

Worked Example

Some preparation of the reference material is often required before it can be introduced to the instrument and frequently the preparation of a solution of the calibration standard is necessary. Such operations will, in principle, add to the uncertainty of the documented property value. The following example, using LGC CRM1110 shows how the preparation of a solution can affect the uncertainty of the property value.

LGC 1110, which is p,p'-DDE with a certified purity value of 99.6%, has an expanded uncertainty (coverage factor, k=2) of 0.4%. A solution of the material, in hexane, with a nominal concentration of 0.1mg/mL was prepared and Table 5 shows the values for the purity of the CRM, the mass of the CRM taken and the volume of hexane solvent used, together with their estimated standard uncertainties.

Table 5

PREPARATION OF A CALIBRATION SOLUTION USING A CRM (1)					
Source of uncertainty	**Units**	**Value**	**u**	**ru**	**ru2**
certified purity		0.996	0.002	0.002008	4.03E-06
mass of CRM taken	g	0.0114	0.0002	0.017544	0.000308
volume of hexane	mL	100	0.06	0.0006	3.6E-07
Concentration	**mg/mL**	**0.1135**	**0.0020**	0.017669	0.000312
u = standard uncertainty ru = relative standard uncertainty (=u/value) ru2 = relative standard uncertainty squared					

The concentration of the prepared solution is calculated to be 0.1135 mg/mL. The standard uncertainty of this value, calculated by combining the individual uncertainties in the usual manner, is seen to be 0.0020 mg/mL, which is equivalent to 1.77% of the concentration value. Thus the use of this solution for an instrument calibration would result in a calibration with a standard uncertainty of at least this value. It is of interest to note from the column "ru" in Table 5 that the mass value makes by far the largest contribution to the standard uncertainty of the concentration value. The original CRM has a standard uncertainty that is only 0.20% of the certified purity value. Thus the preparation of the solution has introduced significant additional uncertainty over and above that due to the certified reference material as issued by the producer. The analyst would therefore need to decide whether the standard uncertainty of the prepared solution of 1.77% was acceptable for a particular analytical application.

If the standard uncertainty of 1.77% is not acceptable, the solution would need to be prepared in a different manner. One simple approach would be to carry out the weighing of the CRM using a 5-figure, rather than a 4-figure, balance. The uncertainty associated with the mass value of 0.01143g is then estimated to be 0.00004g , which leads to a standard uncertainty of 0.00046mg/mL for the solution concentration, which is 0.40% of the concentration value, as shown in Table 6.

Table 6

PREPARATION OF A CALIBRATION SOLUTION USING A CRM (2)					
Source of uncertainty	**Units**	**Value**	**u**	**ru**	**ru2**
certified purity		0.996	0.002	0.002008	4.03E-06
mass of CRM taken	g	0.01143	0.00004	0.0035	1.22E-05
volume of hexane	mL	100	0.06	0.0006	3.6E-07
Concentration	**mg/mL**	**0.11384**	**0.00046**	0.004079	1.66E-05
u = standard uncertainty ru = relative standard uncertainty (=u/value) ru2 = relative standard uncertainty squared					

Although inspection of Tables 5 and 6 suggests that the measured volume of the hexane solvent is a negligible contributor to the standard uncertainty of the solution concentration, it will be realised, on further consideration, that the ambient temperature may influence the solution volume and therefore its concentration. For example, if the solution was prepared at 20°C, but then subsequently used in a laboratory where the ambient temperature was 25°C, the expansion of the hexane solvent will mean that the actual concentration of the calibrant as introduced to the instrument is less than the prepared value (0.1135mg/mL in the example). Based on knowledge of the thermal expansion of hexane, it can be calculated that 100.0mL of hexane at 20°C, will occupy 99.3mL at 15°C and 100.7mL at 25°C. Thus, if we are confident that we can control the temperature of the calibration solution during both preparation and use to 20 ±5°C, then the standard uncertainty in the hexane volume may be estimated (assuming a rectangular distribution) as $0.7/\sqrt{3}$ = 0.40mL. Comparison of this estimate for the standard uncertainty of the hexane volume with that given in Table 5 of 0.06mL, shows the influence of temperature effects on standard solutions in organic solvents. Table 6 shows the effect on the uncertainty of the concentration value. For example, if the ambient temperature varies by ±5°C during the preparation and use of the calibration solution, the standard uncertainty of the concentration value increases from 0.00046 mg/mL to 0.00065 mg/mL.

Table 7

PREPARATION OF A CALIBRATION SOLUTION USING A CRM (3)					
Source of uncertainty	**Units**	**Value**	**u**	**ru**	**ru2**
certified purity		0.996	0.002	0.002008	4.03E-06
mass of CRM taken	g	0.01143	0.00004	0.0035	1.22E-05
volume of hexane	mL	100	0.4	0.004	0.000016
Concentration	**mg/mL**	**0.11384**	**0.00065**	0.005681	3.23E-05
u = standard uncertainty ru = relative standard uncertainty (=u/value) ru2 = relative standard uncertainty squared					

The general conclusion is that the analyst needs to be aware and take account of any aspects of the procedure used to prepare the calibration solution that may adversely affect the documented properties of the original reference material.

Of course, it should be appreciated that the above considerations are entirely different from the situation where the solvent actually evaporates from the solution, causing a gradual increase in the concentration value. The introduction of uncontrolled systematic errors of this type must be avoided by proper storage and possibly by appropriate monitoring (e.g. total mass measurements of the solution plus container) of the calibration solution.

Quite often a "stock" calibration solution will be prepared and from this a number of more dilute solutions will be prepared, so as to cover the range of analyte concentrations of interest to the analyst. Such dilutions will contribute to the uncertainty of the (diluted) concentration value. Table 8 illustrates the nature of these effects for the situation where an aliquot volume of 1mL of the stock solution, measured by pipette, is diluted to 10mL in a volumetric flask.

Table 8

PREPARATION OF A CALIBRATION SOLUTION USING A CRM (4)					
Source of uncertainty	Units	Value	u	ru	ru2
stock solution	mg/mL	0.11384	0.00046	0.004041	1.63E-05
aliquot volume(pipette)	mL	1	0.0056	0.0056	3.14E-05
final volume (flask)	mL	10	0.0144	0.00144	2.07E-06
Diluted concentration	**mg/mL**	**0.011384**	**8.03E-05**	0.007054	4.98E-05
u = standard uncertainty ru = relative standard uncertainty (=u/value) ru2 = relative standard uncertainty squared					

Inspection of the column "ru" in Table 8 shows that the standard uncertainty of the diluted solution is 0.7% of the concentration value, compared to 0.4% for the undiluted stock solution. The analyst will need to consider the significance of such effects for the particular measurement he intends to carry out since the analyical result on a sample cannot be any better (and will be probably be worse) than the accuracy of the calibration standard. Furthermore, it should be noted that the use of the statistical technique of linear regression to construct the best straight line through a set of calibration points depends on the assumption that the uncertainty in the concentration values of the calibration standards is insignificant compared to the uncertainty in the measured instrument signal for those standards. A knowledge of the expected uncertainties in the calibration solution concentrations allows the analyst to assess the validity of this assumption for a particular set of calibration data.

Table 9

PREPARATION OF A CALIBRATION SOLUTION USING A CRM (5)					
Source of uncertainty	Units	Value	u	ru	ru2
certified purity		0.996	0.002	0.002008	4.03E-06
mass of CRM taken	g	0.11431	0.00004	0.00035	1.22E-07
mass of hexane	g	66.1054	0.0005	7.56E-06	5.72E-11
Stock concentration	mg/g	1.7223	0.00351	0.002038	4.15E-06
aliquot mass	g	0.65892	0.00004	6.07E-05	3.69E-09
final mass	g	65.9046	0.0005	7.59E-06	5.76E-11
Diluted concentration	**mg/g**	**0.01722**	**3.51E-05**	0.002039	4.16E-06
u = standard uncertainty ru = relative standard uncertainty (=u/value) ru2 = relative standard uncertainty squared					

If the uncertainty in the prepared calibration solutions is considered to be too high, one option would be to prepare the solutions entirely by mass. Table 9 shows the calculated uncertainties for where the stock solution was prepared by taking a nominal 0.1g of the CRM and dissolving it in a nominal 66g of hexane

($\approx$100mL). A nominal 0.66g aliquot of the stock solution was then diluted to a nominal 66g with hexane. It is seen from the column "ru" in Table 8 that the standard uncertainty of the diluted calibration solution is 0.20% of the concentration value. This compares with a standard uncertainty that was 0.7% of the concentration value when the dilution was carried out by volume. It may also be noted that for the dilution by mass, the standard uncertainty of 0.20% is due entirely to the standard uncertainty of the certified purity value of the CRM itself. Thus, by preparing the calibration solution in the manner shown in Table 9, no significant addition to the uncertainty of the original CRM has been incurred.

8 SUMMARY

The correct use of reference materials in a wide variety of analytical measurements (calibration, validation, quality control, etc.) is crucial to obtaining results which are fit for purpose. This paper has described some of the more common applications of reference materials using a number of worked examples to illustrate the principles involved.

References

1. ISO Guide 31 (1981), ISO, Geneva.

2. ISO Guide 34 (1996), ISO, Geneva.

3. ISO Guide 35, 2nd edition (1989), ISO, Geneva.

Acknowledgements

Preparation of this paper was supported under contract with the UK's Department of Trade and Industry as part of the National Measurement System's Valid Analytical Measurement (VAM) Programme.

Are We Clear on the Function of Matrix Reference Materials in the Measurement Process?

P. De Bièvre

INSTITUTE FOR REFERENCE MATERIALS AND MEASUREMENTS, EUROPEAN
COMMISSION – JRC, B-2440 GEEL, BELGIUM

1 INTRODUCTION: THE AIM OF A MEASUREMENT

Measurement processes must be described in terms of the quantity which we want to describe quantitatively. Thus we can want to do a measurement of the quantity 'time', or of the quantity 'temperature', or of the quantity 'electric current'. In chemical measurement, we may want to do a measurement of the quantity 'amount (of substance)', or of a derived quantity such as concentration (amount per volume), or content (amount per mass), or mass fraction (mass per mass). Provided our result is stated in terms of the quantity announced a priori and hence in the internationally agreed units for that quantity, that is OK. A problem only arises when we intend to measure one quantity, yet report or claim to have measured another quantity. This frequently occurs in chemical measurement: an amount is claimed, or at least implied, and mass is used. Both are useful, but internal consistency and compatibility in executing and reporting the measurement is to be strongly recommended. We don't announce a temperature, then report the result in ampère, although we measure the electric current generated by a thermocouple. Therefore, in chemical measurement, we report an amount, an amount per volume (= concentration), an amount per mass (= content), mass per mass (= mass fraction).

   Another basic feature of our "measuring" is, that we do not in fact measure "reality", but *the model of reality we have in our mind*. When we measure, we do not check whether our quantitative observations correspond with reality itself, but with our model of reality which is our way of expressing (one of) our best perception(s) of reality so far. The degree of correspondence between our measurements and our best model of reality, can then be quantified (best fit of data, systematic deviations, etc). The degree to which we can quantitatively approach such a model, is called "uncertainty". It can be evaluated from how well our measurements confirm our model of reality. Thus any discussion about "true value", if meant to pertain to "reality", is meaningless. Consequently, all ISO definitions using the concept - or being based on the concept - of "true value", are also meaningless, because not applicable in practice .We will not use them anymore. They should be eliminated from the International Vocabulary for Basic and General Terms in Metrology (VIM) and ISO Guides and Standards altogether.(They have already been eliminated from the ISO Guide on the Expression of Uncertainty in Measurement (GUM).[1]

It is useful to bring the above basic clarifications before starting to discuss the question on the function of matrix reference materials in the measurement process. Before we do that, we need to be clear on yet another basic concept.

2 WHAT IS A CALIBRATION?

According to the VIM, a calibration is "the set of operations that establish, under specified conditions, the relationship between values of quantities indicated by a measuring instrument or measuring system, or values represented by a material measure or a reference material, and the corresponding values realized by standards".[2] Applying that to the case of a chemical measurement, we have concluded previously[3] that a calibration should be "the process by which a quantitative relation over a range of observed responses is established, correlating each of several known concentrations to its corresponding signal, thus yielding a response curve". This wording is very much influenced by the original practical meaning of the word, as well as by daily work and needs in the laboratory. Figure 1 illustrates what we mean by that. It is applicable fairly generally: it correlates each of several known concentrations ("the corresponding values realized by standards") to its corresponding signal, i.e. the observed response from an instrument + measurement procedure (" ... values of quantities indicated by a measuring instrument or measuring system ... ").

CALIBRATION:
the process by which a quantitative
relation over a range of observed responses
is established correlating each of several known concentrations to
its corresponding signal, thus yielding a response curve

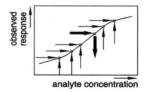

analyte concentration

———▶ establishing the calibration curve
━━━▶ using the calibration curve to measure the value of an inknown

Note: The reverse of this function is sometimes called an "analytical function"

Figure 1 *Calibration: the process by which a quantitative relation over a range of observed responses is established, correlating each of several known concentrations to its corresponding signal, thus yielding a response curve.*[1]

There is an important characteristic involved in a calibration process as described above: the conversion of what we actually measure to what we purport (claim) to have measured. We almost never measure what we claim at the end of a chemical measurement to have measured. We measure light absorption, or light emission, which then is converted to an electric current that is measured .At the end of the process we claim to have measured a concentration. We measure ions in the liquid or the gas phase and measure electric currents, but again claim to have measured concentration. There are a number of steps in the conversion of what we actually measure (e.g. electric current) to what we claim to have measured (e.g. concentration), which we do not (yet) fully understand and which, therefore, carry known and unknown uncertainties.   It is these conversion steps - all uncertain to some degree - which the measurement laboratory can spare if it has a means for "calibration". It makes quick, "calibrated" measurements possible by leaving the burden of full understanding of the "values realized by standards" to a (national) measurement institute or metrology laboratory.

Thus in Figure 1 we observe that the "known" points on the abscissa can be the (certified) values carried by (matrix) reference materials. This simple picture enables us

    a) to understand a "calibration process" in a clear, transparent way

    b) to "see" the function of matrix reference materials

    c) to conclude that a 'set of reference materials' are essential to a calibration process

    d) to observe that this set can be certified in any unit which is convenient

    e) to understand that the unit to express the measurement result (and uncertainty) for the unknown, will automatically be the same as the unit chosen to certify the values of the matrix reference material(s)

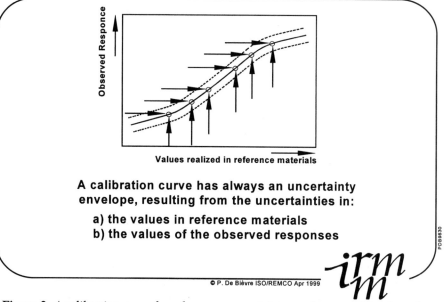

A calibration curve has always an uncertainty envelope, resulting from the uncertainties in:

a) the values in reference materials
b) the values of the observed responses

© P. De Bièvre ISO/REMCO Apr 1999

Figure 2 *A calibration curve has always an uncertainty envelope, resulting from the uncertainties in: a) the values in the reference materials used, and, b) the values of the observed responses*

We note that this original and transparent meaning of "calibration" is somewhat lost and that the term "calibration" is used (and misused) in many contexts. But the introduction of more metrology-oriented thinking in chemical measurement requires a more unequivocal use of terms. We propose that "calibration" be reserved for use as described above.

There is more to be learned from Figure 1: uncertainty. That is made explicit in Figure 2.

3 UNCERTAINTY OF A MEASUREMENT USING (A) MATRIX REFERENCE MATERIAL(S) IN CALIBRATION

We note that a "calibrated" measurement carries an "uncertainty" component generated by each of the conversion steps in the measurement process. Thus there must be a numerical uncertainty associated with any "overall calibration".

Figure 2 enables us to see that:

a) any calibration point on the curve has an uncertainty, composed by the uncertainties of BOTH abscissa and ordinate values; the former must be taken from the *certificate of the reference material*, the latter necessarily results from the *measurement of the reference material*

b) the uncertainty of the result of a measurement will, of necessity, be composed of the uncertainty of the calibration point on the curve which is used (see above) PLUS the uncertainty (usually the repeatability or the reproducibility) of the measurement of the unknown

An important consequence of these uncertainty considerations, would be that not one, but several points are necessary to establish a calibration curve in order to make possible a full uncertainty assessment of a measurement of an unknown. The shape of the curve might be linear in the neighbourhood of the value to be measured That can only be estimated by at the very least two (for a linear curve), possibly three (for a non-linear curve), preferably four to five points (idem). It is possible to use only one point - as mostly done in practice - , but the price to be paid for having only one point "determining the calibration curve", is simply a (much) larger uncertainty in the measurement result. **The desired uncertainty of the result - or Target Value for uncertainty - determines the size of the effort to be put in the measurement, part of which is the calibration.** It is automatic that quick, overall calibrations by means of one reference material only, must lead, by definition, to larger uncertainties than if carried out using a set of reference materials bringing different values and, hence, a better knowledge of the calibration curve. Consequently, setting a "target value" for uncertainty prior to a measurement[4,5] is very important: it determines the size of the effort to be spent in the calibration and therefore of the size of the cost of the measurement.

We now turn our attention to "traceability".

4 TRACEABILITY OF THE MEASUREMENT RESULT

Figure 3 has been developed in order to understand to what the result of a "calibrated" measurement of an unknown is "traceable": to the measurement scale, (in whatever units), created by the values carried by the set of reference materials. *Traceability to a measurement scale* has thus been realized in a transparent way. It is obvious from the definition - but often overlooked - that traceability is a characteristic of the measurement result ie of a value. Indeed, the VIM definition of traceability is [2]: "property of the result of a measurement or the value of a standard whereby it can be related to stated references, usually national or international standards, through an unbroken chain of comparisons all having stated uncertainties". We conclude that the requirements of this definition are met in Figure 3.

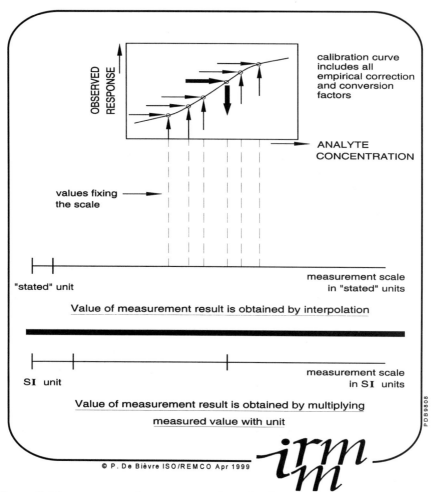

Figure 3 *Measurement scales can act as "stated references" for the establishment of traceability*

How does all of this relate to "traceability to the SI" of amount of substance measurements? By virtue of the definition, the expression "measurements of amount of substance" is only applicable to well-identified substances. That is: substances which can be identified, then counted. We need to remember that the quantity "amount of substance" and its unit "mol", were consecrated into a "base quantity" resp a "base unit" in 1971, at the explicit request of the chemical community. The concept of amount of substance and its unit mole (symbol: mol), reflect our recognition of the "particulate nature of matter" and the fact that in chemical reactions, species interact in ratios of numbers. It is essential to understand that amount relates to a "number of things" (the formal word is "entities") which need to be well identified (atoms, molecules of well specified substances, but also electrons, ...). That enables to conceive of a measurement scale along which a "count" of such entities can be made. Since that would involve counting very large numbers, an already large number has been chosen as unit: a mole. Thus unknown large numbers can be measured - ratio-ed - against a known large number defining the unit[6] This can be visualized in the form of a simple measurement scale as illustrated in Figure 4.

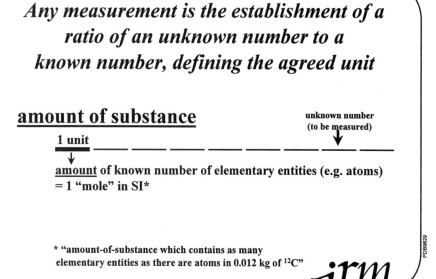

Figure 4 *SI measurement scale for an amount-of-substance measurement*

The similarity of such a scale as compared to a measurement scale derived from a "calibration", is striking as shown in Figure 3. It looks as if the SI measurement scale with a unit mol, is a particular case of a measurement scale in general (with whatever unit). Thus, we propose that traceability of chemical measurement be expressed to values on a measurement scale with a specified unit. The similarity of scales for measuring amount of well specified and less well specified substances is striking and simple. It follows naturally that measurements can be made traceable to (the scale with) the SI unit or to an empirical scale with an empirical (yet useful) unit. The use of measurement scales, as proposed here, is of course not new, but is clearly not implanted anymore in the mind of chemists. It must also be said that the introduction of "quantity calculus", mainly in so-called "physical" measurements (more about that in ref. 7), obviated the need for a "measurement scale", contributing to its going out of use in the so-called "physical measurements" and to its being unknown in the world of "chemical measurements". (In quantity calculus, a measurement value for a quantity results from the product of a numerical value times the unit for that quantity). Therefore, the proposal to illustrate the "realization of traceability" by means of measurement scales, is meant to be helpful and "educational" in the first place.

5 CONCLUSIONS

Matrix reference materials can play an essential role in establishing measurement scales with a "stated unit" which seems to be an excellent and simple way to realize traceability of measurement results to a "stated reference", not necessarily to the SI. They also help to realize the badly needed calibration curves which are extremely useful tools to convert what we actually measure, to what we purport (claim) to have measured, at the same time enabling an "overall correction factor" for a number of (known and /or unknown) systematic errors.

But, also, they are very helpful in that they bring forcefully to our attention the very important need to do proper uncertainty assessment, including every part of the measurement process, also the calibration!

Understanding the role of matrix reference materials in a world where a huge number of chemical measurements are made daily in a variety of matrices, is useful - because it enables to optimize the use of these reference materials in the measurement process in relation to their fitness for purpose. It is also cost-effective, because it enables to see that a price has to be paid for a given quality ie for a required (total) uncertainty.

References

1. Guide on the Expression of Uncertainty in Measurement, BIPM, IEC, IFCC, ISO, IUPAC, IUPAP, OIML, ISO Genève 1993, 1st Edition
2. International Vocabulary of basic and general Terms in Metrology, BIPM, IEC, IFCC, ISO, IUPAC, IUPAP, OIML, ISO Genève 1993, 2nd Edition
3. P. De Bièvre, R. Kaarls, H. S. Peiser, S. D. Rasberry and W. P. Reed, *Accred Qual Assur* 1997, **2**, 170,
4. P. De Bièvre, *Accred Qual Assur* 1998, **3**, 139
5. P. De Bièvre, *Accred Qual Assur* 1998, **3**, 179,
6. P. De Bièvre, S. Valkiers and P. D. P. Taylor, *Fresenius J Anal Chem* 1998 **361** 227-234
7. J. de Boer, Metrologia, 1994/5 **32** 405

Uses of Certified Reference Materials: The ISO Guide 33

Eva Deák

DIVISION OF CHEMISTRY, NATIONAL OFFICE OF MEASURES, BUDAPEST, HUNGARY

1 INTRODUCTION

It is evident, that the demand is increasing worldwide for uniform, compatible and reliable chemical measurements. The global harmonization of analyses is important both for environment controlling, healthcare, safety and for improvement the quality of industrial products. Nowadays more and more accredited laboratories make efforts to assure the traceability and comparability of own results. At a great number of cases the best – if not the only – way to reach this aim is the use of reference materials.

The estimated number of existing kind of reference materials approximately thirty thousand, the producers are a few hundred, the users are countless and they are going to be more and more. To reach compatibility of measurement results under such conditions it is essential that the reference materials themselves, originated from different sources, should be equivalent and their use should be uniformly regulated. From the late seventies different international organizations try to response this challenge creating guides, proposals, standards and other regulations. During the years a committee of the International Organization for Standardization, ISO/REMCO became the main authority worldwide. ISO/REMCO, summarizing the experiences of different specialists from many countries, 59 member countries at present, and collaborating with 12 other international organizations, provides guidance to the preparation and the proper use of reference materials.

Achievements of ISO/REMCO are reflected in the Guides published by ISO. ISO/REMCO has begun his work in logical order, first of all the terminology and definitions were stated to avoid ambiguity in discussing the following tasks: to give guides for the certification, for the use and for the production of certified reference materials. The first two guides, ISO Guide 30: *Terms and definitions used in connection with reference materials* and ISO Guide 31: *Contents of certificates of reference materials* were published in 1981 (first editions). After long and serious discussions the original different contributions of the specialists were edited and two further guide were published in 80`s: ISO Guide 33: *Uses of certified reference materials* and ISO Guide 35: *Certification of reference materials - General and statistical principles*.

The alteration in terminology, the widespread application of certified reference materials in the last decade and especially the publication of the Guide[1] to the Expression of Uncertainty in Measurement made necessary the revision of these ISO/REMCO documents. The new, corrected version of Guides 30, 31, 32 and 33 were published (or under publication) recently[2-5]. Guide 35[6] is under revision this year.

Guide 32 published in 1997 provides principles how to use reference materials for calibration in analytical chemistry. ISO 11095:1994 standard[7] gives a statistical guidance for linear calibration using reference materials. As a consequence the new edition of Guide 33 does not deal with this important field of application of certified reference materials only refers the mentioned documents.

The purpose of this paper is the familiarization with Guide 33 pointing out the proper use of certified reference materials.

2 DEFINITIONS

For the better understanding of the use of certified reference materials at least two definitions have to keep in mind. By wording of international metrological vocabulary[8] these sound as follows:

reference material (RM) : A material or substance one or more of whose property values are sufficiently homogeneous and well established to be used for the calibration of an apparatus, the assessment of a measurement method, or for assigning values to materials.

certified reference material (CRM) : A reference material, accompanied by a certificate, one or more of whose property values are certified by a procedure which establishes traceability to an accurate realization of the unit in which the property values are expressed, and for which each certified value is accompanied by an uncertainty at a stated level of confidence.

3 THE ROLE OF CRMS IN MEASUREMENT SCIENCE

3.1 Dependence of the SI Base Units on Substances and Materials.

The majority of measurements made in the world today are within the framework of the International System of units. In its present form, SI recognizes seven base units, namely the units of length (metre, symbol m), mass (kilogram, kg), time (second, s), electric current (ampere, A), thermodynamic temperature (kelvin, K), amount of substance (mole, mol) and luminous intensity (candela, cd). The definitions of these base units mention the following substances: platinum-iridium (for fabricating the prototype kilogram), caesium-133 (for the second), water (for defining the kelvin) and carbon-12 (for defining the mole).

Opinions differ as to whether the substances named fall under the definition of reference material. Certainly such materials have a special status as defined substances on which the SI is based. The dependency strictly applies to definition of the unit, since realization of the units may involve other substances/materials.

3.2. The Realization of Derived SI Units with the Aid of Reference Materials.

From the seven base units of the SI a large number of derived units of the SI are obtainable by combining base units as products and/or quotients. For example, a derived unit of mass concentration is defined as $kg\ m^{-3}$ and the derived unit of pressure (given the special name pascal, symbol Pa) is defined as $m^{-1}\ kg\ s^{-2}$. Formally speaking, the derived units ultimately depend on the substances on which the base units themselves depend (see 3.1). In practice, the derived units are often realized not from base units but from CRMs with accepted property values. Thus a variety of substances/materials may be involved in the realization of derived units (Examples 1 and 2 below) or even of base units (Example 3 below).

Example 1: The SI unit of dynamic viscosity, the pascal second (Pa s = $m^{-1}\ kg\ s^{-1}$) may be realized by taking the value for a well purified sample of water as 0.001002 Pa s at 20°C.

Example 2: The SI unit of molar heat capacity, the joule per mole kelvin
($J\ mol^{-1}\ K^{-1}$ = $kg\ m^2\ s^{-2}\ mol^{-1}\ K^{-1}$) may be realized by taking the value for purified α-alumina as 79,01 $J\ mol^{-1}\ K^{-1}$ at 25°C.

Example 3: The SI unit of temperature, the kelvin, may be realized at any temperature T_1 (273,15 K < T_1 < 903,89 K) from measurements of the resistance of a highly pure platinum wire at the triple point of purified water, at the freezing point of purified tin and at the freezing point of purified zinc, coupled with use of a specified mathematical relation.

3.3 The Role of Certified Reference Materials in the Storage and Transfer of Information on Property Values.

By definition, a reference material has one or more properties, the values of which are well established. Once the property value(s) of a particular CRM have been established, they are "stored" by the CRM (up to its expiration date) and are transferred when the CRM itself is conveyed from one place to another. To the extent that the property value of a CRM can be determined with a well-defined uncertainty, that property value can be used as a reference value for intercomparison or transfer purposes. Hence CRMs aid in measurement transfer, in time and space, similar to measuring instruments and material measures.

A CRM must be suitable for the exacting role it performs in storing and transferring information on measured property values. Strict technical criteria (legal or commercial criteria may be relevant also) apply to the fitness for purpose of CRMs; these will be treated later in the paper.

Whenever possible, the measurement of a given property value should have been made by an acceptable method having negligible uncertainty relative to end-use requirements and by means of measuring instruments or material measures which are traceable to national measurement standards. Subsequent use of a CRM with traceable property values ensures that traceability is propagated to the user. Since most national measurement standards are themselves harmonized internationally, it follows that measurement standards in one country should be compatible with similar measurements in another country. In many cases, CRMs are appropriate for the intercomparisons of national measurement standards.

3.4 Defining and Realizing Conventional Scales

Many measurement scales have been used since the earliest civilizations. Originally almost all of them were conventional, independent and inaccurate. Scientific and technical progress as well as international trade have brought both the need and the possibility of a unique, rational, self-consistent international system of units, the SI, which has been officially adopted worldwide. Nevertheless, it is not applicable to certain types of measurements for which it is necessary to create, sustain and use certain conventional units which are not within the scope of SI. In other cases the unit relating to the quantity to be measured lies within the frame of SI, but the reproduction of the unit according to the definition is technically difficult and expensive. The realization of the measurement is therefore more convenient on a practical scale of reference values assigned to material properties. Though a reference value scale and a pure conventional scale differ theoretically from each other, they are similar with respect to the use of reference materials, and they are therefore be discussed together as conventional scales.

Conventional scales are based on the values assigned to reference materials. The assigned values are stated in standard specifications, international recommendations or other reference documents; therefore a reference material realizing a fixed point on a conventional scale should have the same quality all over the world. CRMs of this type are certified for property values, i.e. they are measured on standard equipment with reference methods at metrological or other authorized laboratories.

It is evident that the CRMs ensure only the fixed points of a measurement scale. Measurement on a scale requires either a fixed point and a mathematical function passing through it, or two or more fixed points with stated means of interpolation between them.

A conventional scale has two fundamental pillars : the certified reference material, realizing the fixed point(s), and the standard specification (or similar document), giving the method of measurement. Both of them should be strictly defined to ensure the compatibility of measurements on the conventional scale.

The standard specification provides detailed information necessary to establish and use a scale based on assigned values or it may provide protocols for the experimental and calculational procedures to be used in measurements which depend on assumptions. It is advisable to prescribe the requirements of the certified reference material in the same standard specification as that in which the method of measurement is described. By means of the necessary CRMs and relevant standard specifications, the user can realize the measurement scale and with the aid of such a scale can measure his sample or calibrate his instrument.

To estimate the uncertainty of a measurement on the scale, the user should consider the uncertainties in the creation of the scale and the uncertainty associated with the realization of its fixed points by the CRM. Sometimes the users demand a level of uncertainty in the end-use which is lower than the uncertainty of the fixed points defined by the CRM (e.g. in measurement of the pH of blood). They need to realize that the uncertainty of the measurements on the scale is necessarily greater than the uncertainty of the fixed points. The replicated measurement of the CRM, and the setting-up of a scale (the appropriate selection of the points, the characteristics and repeatability of the interpolating instrument, etc.) also contributes to the overall uncertainty.

The selection of CRMs for realizing the fixed points of a scale should be directed by the required level of uncertainty of the end use. To minimize the uncertainty of the measured value on the scale, the user should employ CRMs which have been certified in terms of the units of the scale. Obviously, the user is expected to be familiar with all relevant information about the method for realizing the scale and the instructions for the correct use of CRM.

In certain cases the user can apply pure chemical compounds for realizing the fixed points if CRMs certified in the scale units are unavailable or expensive, or if their use is not necessary at the level of the uncertainty of the measurement. If this method is chosen, the user should be aware of the correlation between the purity of the material and the property on which the scale is based, and the uncertainty of the measurement can be only roughly estimated.

There is a great variety of conventional scales and the methods of application of the CRMs for realizing them differ widely. Some examples illustrating widely used conventional scales are: the International Temperature Scale (ITS-90), the pH-scale, the octane-number scale, turbidity scale.

4 PROPER USE OF CRMS IN LABORATORY PRACTICE

CRMs are very precious measuring instruments; metrologically they are measurement standards. They are too expensive for careless use and, which is more important, the misuse, the incorrect use, of CRMs may not provide the intended information, may cause misleading even harmful results, bad decisions.

A potential CRM user should prepare first of all his laboratory operating an internal quality system. This is a pre-requisition, without any statistical control it is no sense to use CRM.

The next stage is the accurate determination of the intended use of CRM. The further considerations will depend on this decision. The certified reference materials can be applied:
- for calibration of an apparatus
- for establishing traceability of the measurement result
- for determining uncertainty of these traceable results
- for assessment of a measurement method

A CRM should not be used for a purpose other than that for which it was intended. Nevertheless, from time to time, when a user must resort to applying a CRM in an incorrect manner because of the unavailability of a suitable CRM, he must be fully aware of the potential pitfalls and therefore assess his measurement output accordingly.

The third step what the potential CRM user has to do: the choice of the CRM which fit for the purpose. At this point the user should study all the information provided by the producer, the certificate and the instruction of use would be scrutinized thoroughly. It is essential to read and keep in mind all the information that accompanies the product. Attention should be paid such points as homogeneity, stability, best storage conditions for an optimal shelf life. For certain CRMs, the level of homogeneity is valid for a test portion defined by mass, physical dimension, time of measurement, etc. The user must be aware that the use of a test portion that does not meet or exceed that specification could severely increase the contribution of the inhomogeneity of the CRM to the uncertainty of the certified property to the point where the statistical parameters of certification are no longer valid.

The user of the CRM must decide what properties of the CRM are relevant to his measurement procedure, taking into account the method of certification, the statement on intended use and instructions for the correct use of the CRM on the certificate. The following properties have to be considered:

a) Level. The CRM should have properties at the level appropriate to the level at which the measurement process is intended to be used, e.g. concentration.

b) Matrix. The CRM should have a matrix as close as possible to the matrix of the material to be subjected to the measurement process, e.g. carbon in low-alloy steel or carbon in stainless steel.

c) Form. The CRM may be in any physical state and form, e.g. solid, gas, etc. It may be a test piece or a manufactured article or a powder. It may need preparation. It must be used in the same form as the as the sample to be measured.

d) Quantity. The quantity of the CRM must be sufficient for the entire experimental programme, including some reserve if it is considered necessary. Avoid having to obtain additional new batches of the CRM later in a given measuring process.

e) Stability. Wherever possible the CRM should have stable properties throughout the experiment. There are three cases :
- the properties are stable and no precaution is necessary;
- when the certified value may be influenced by storage conditions, the container should be stored, both before and after its opening, in the way described on the certificate;
- a certificate is supplied with the CRM defining the properties (which are changing at a known rate) at specific times.

f) Acceptable uncertainty of the certified value. One of the important considerations in selecting a CRM for use either in assessing the trueness and precision of a method or in the calibration of instruments in a method is the level of uncertainty required by the end-use of the method. The uncertainty of the certified value should be compatible with the precision and trueness requirements. Obviously the user should not apply a CRM of greater uncertainty than permitted by the end-use.

The selection of CRMs must take into account not only the level of uncertainty required for the intended purpose but also their availability, and chemical and physical suitability for the intended purpose. For example, the unavailability of one CRM could force a user to resort to using another CRM of greater uncertainty than the preferred one. Also, in chemical analysis, a CRM of greater, but still acceptable, uncertainty in the certified property may be preferred over another CRM because of better matching with the composition of real samples. This could result in minimizing "matrix" or chemical effects in the measurement process which are capable of causing errors far greater than the difference between the uncertainties of the CRMs.

g) Method of certification. For CRMs certified by a primary method, the user should not assume that his method is capable of matching the precision and trueness reported for the CRM. It is unreasonable therefore to apply the statistical procedures given in Guide 33 for assessing the trueness and precision of a method by application to a CRM using the certification parameters for a property reported in the certificate. The user, as a consequence, must either experimentally establish or make estimates based on available information for those parameters that are more appropriate. Similarly, where a user applies a method to a CRM that has been certified by a single different method, the user must not assume that the certification parameters for the certified property are applicable to his method except in cases where the trueness and precision capable by both methods are known to be comparable.

5 ASSESSMENT OF A MEASUREMENT PROCEDURE

The procedure for the measurement must be fixed, i.e. a written document must exist laying down all the details. There must be no changes to the procedure during the course of the experiment.

Guide 33 provides the necessary statistical functions to the evaluation of precision and trueness of the measurement method. The criteria used for assessment of a measurement process described in Guide 33 are the prescribed limits for both precision and bias. In order that these criteria are workable, they must be compatible both with the CRM and the state of the art of the measurement process.

5.1 Statistical Considerations

5.1.1 Basic assumptions. All statistical methods used in Guide 33 are based on the following assumptions.

a) The certified value is the best estimate of the true value of the property of the CRM.

b) All variation, be it associated with the material (i.e. homogeneity) or the measurement process, is random and follows a normal probability distribution. The values of probabilities stated in Guide 33 assume normality. They may be different if there is deviation from normality.

5.1.2 Decision errors. The assessment of a measurement process on the basis of precision and trueness is always subject to rendering an incorrect conclusion because of

a) the uncertainty of measurement results and

b) the limited number of replicate results usually performed.

Increasing the number of measurements tends to decrease the chance of an incorrect conclusion but, in many instances, the risk of making a wrong conclusion has to be balanced in economic terms against the cost of increasing the number of measurements. Accordingly, the rigour of the criteria developed for assessing a measurement process must take into account the level of precision and trueness requisite for the end-use.

For the purposes of Guide 33 the term "null hypothesis" is applied. In this case the null hypothesis is that the measurement process has bias no greater than the limit chosen by the experimenter and variance no greater than the predetermined value; the alternative hypothesis is the hypothesis which is opposed to the null hypothesis (see ISO 3534-1[9]).

5.2 Requirements of Limits

In order to satisfy the requirement, the measurement procedure must produce results with a precision measure and/or trueness within the predetermined limits when it is applied to a CRM. The limit of precision is usually expressed in terms of standard deviation and the trueness requirement is expressed in terms of the bias of the measurement results against the certified value. These limits may originate from various sources.

- Legal limits. Legal limits are those limits which are required by statute or regulation; for example, procedures for the analysis of sulphur 'dioxide in air are required in many countries to have a certain precision and trueness.

- Limits in accreditation (proficiency testing) schemes. In most cases the limits of uncertainty are consensus values agreed upon between the various participants concerned, e.g. producer, consumer and independent. For this reason these limits usually are derived from some realistic values, e.g. those obtained from the results of the certification campaign of the CRM, international tests of ISO standards, etc.

- Limits given by the user of the procedure. In this case the laboratory, or the organization of which the laboratory is a part, imposes upon itself the limits of bias and precision, e.g. limits imposed by commercial requirements.

- Limits from previous experience. In this case the limits of bias and precision of the measurement process to be tested are based on the values obtained from previously established measurement processes.

5.3 Check of Precision and Trueness of a Measurement Process by One Laboratory.

One particular laboratory uses a CRM to check its measurement performance for any particular reason at any time.

Checking of precision of a measurement procedure as applied by a laboratory involves comparison of the within laboratory standard deviation under repeatability conditions (or other defined conditions) and the required value of standard deviation.

Checking of trueness of a measurement process as applied by a laboratory involves comparison of the mean of the measurement results and the certified value of the CRM. The between-laboratories component of precision of the measurement procedure should be taken into account when making this comparison.

- Measurement. The user should perform independent replicate measurements. "Independent", in a practical sense, means that a replicate result is not influenced by previous replicate results. To perform replicate measurements means to repeat the whole procedure. For example, in the chemical analyses of a solid material, the procedure should be repeated from the weighing of the test portion to the final reading or calculating of the result. Taking aliquots from the same sample solution is not independent replication.

Independent replicate measurements can be achieved in various ways depending on the nature of the process. In some, however, parallel replication is not recommended because an error committed at any step of the procedure could affect all replicates. For example, in the case of iron ore analyses, replication of the analytical procedure is carried out at different times and includes appropriate calibration.

The measurement results could, if necessary, be scrutinized for possible outliers using the rules described in ISO 5725[10]. It should be noted that an excessive number of suspected outliers indicates problems in the measurement process.

- Assessment of precision. The precision of the measurement process is assessed by comparing the within-laboratory standard deviation under repeatability conditions with the required value of the within-laboratory standard deviation.

- Assessment of trueness. The trueness of the measurement process is checked by comparing the average, $\bar{x}$ with the certified value, μ.

There are two factors contributing to the difference between the certified value and the measurement results

1) the uncertainty of the certified value;

2) the uncertainty of the results of the measurement process being assessed expressed by its standard deviation σ_D

For a CRM prepared in accordance with ISO Guide 35[5], the uncertainty of the certified value should be small in comparison with σ_D, the standard deviation associated with the measurement process. The following general condition is used as the criterion for acceptance:

$$-a_2 - 2\sigma_D \leq \bar{x} - \mu \leq a_1 + 2\sigma_D \tag{1}$$

where a_1 and a_2 are adjustment values chosen in advance by the experimenter according to economic or technical limitation or stipulation.

The standard deviation associated with the measurement process, σ_D, arises from the fact that a measurement procedure performed on the same material does not, in general, yield identical results every time it is applied. This fluctuation is attributed to unavoidable random errors inherent in every measurement process because the factors that may influence the outcome of a measurement cannot all be completely controlled. This random fluctuation of the measurement results should be taken into account when assessing the trueness of the procedure..

The mathematical functions and computations are detailed in Guide 33.

5.4 Assessment of a Measurement Process by an Interlaboratory Measurement Programme

One of the most important criteria that a measurement process must satisfy in order to receive "standard" or "widely accepted" status is that it is capable of producing results with precision and trueness sufficient for the end-use when applied by a qualified operator. In most instances, the precision and trueness of such a candidate process are assessed by an interlaboratory measurement programme in which the participants are selected so as to provide a representative sample of the laboratories which will ultimately apply that measurement process. The procedure of conducting an interlaboratory measurement programme is described in ISO 5725[10].

5.4.1 Number of Participant Laboratories, p, and Number of Replicate Measurements per Laboratory, n. Ideally the values of p and n should be selected according to a prescribed limit of bias between the certified value of the CRM and the value obtained by the interlaboratory measurement programme, M, the significance level, α, and other statistical parameters. In many cases, the choice of p and n is limited by the availability of participating laboratories. The detailed procedure for computing the ideal values of p and n is described in Guide 33.

5.4.2 Experiment. An interlaboratory measurement programme is often conducted as part of an experiment to estimate precision of the method. A detailed procedure for performing such an experiment is described in ISO 5725[10].

Check and distribution of the CRM:
a)The suitability of the CRM should be checked.

b)Where subdivision of the unit of the CRM is unavoidable prior to distribution, it must be performed with great care to avoid any additional error. Relevant International Standards on sample division should be consulted. If the unit of the CRM has a fixed form, e.g. a metal disc, the units should be selected on a random basis for distribution. If the measurement process is non-destructive, all laboratories in the interlaboratory measurement programme be may given the same unit of the CRM but this will extend the duration of the programme.

Measurement
The co-ordinator of the interlaboratory measurement programme must specify n, the number of independent replicate measurements to be performed by each laboratory, and the organizational factors of interlaboratory programmes such as time limit for submission of results, the size of test portion, etc.

Methods for computing the precision measures from the results of an interlaboratory programme are described in ISO 5725[10].

Assessment of precision.
The precision of the measurement process as applied to the CRM is expressed in terms of s_w, the estimate of the within-laboratory standard deviation, and s_{Lm} the estimate of the between-laboratories standard deviation.

The χ^2 – test is applied to decide that precision is as good as required.

Assessment of trueness.
The trueness of the measurement process is checked by comparing $\bar{\bar{x}}$, the overall mean of the interlaboratory measurement programme with the certified value of the CRM. In a manner analogous to **5.3** the criterion for acceptance is

$$- a_2 - 2\sigma_D \le \bar{\bar{x}} - \mu \le a_1 + 2\sigma_D \qquad (2)$$

where σ_D is the standard deviation of the overall mean of the interlaboratory comparison for the measurement process, and is given by :

$$\sigma_D^2 = \frac{s_{Lm}^2 + s_w^2 / n}{p} \tag{3}$$

Decision :

1) If condition (2) is satisfied, there is no evidence that the bias of the measurement process exceeds the prescribed limit including the adjustment value.

2) If condition (3) is not satisfied, there is evidence that the bias of the measurement process exceeds the prescribed limit including the adjustment value.

6 CONCLUSIONS

CRMs are meant to fulfil many purposes. The selection of a proper CRM is not an easy task.. The user should study the certificate and the instruction of use and all the information provided by the producer. A CRM used properly for one purpose in one laboratory may be misused for another purpose in another laboratory. It is recommended that the user considers the suitability of a CRM for his intended purpose on a case-by-case basis.

In chemical analysis, a CRM of greater, but still acceptable, uncertainty in the certified property may be preferred over another CRM because of better matching with the composition of real samples. This could result in decreasing matrix effects in the measurement process which are capable of causing errors far greater than the difference between the uncertainties of the CRMs. The determining sources of uncertainty in result are the differences between the composition of CRM and the sample, different operations on CRM and the sample and differences in response of the measuring instrument to CRM and the sample. The result is influenced in a lesser extent by the uncertainties originated from the CRM (uncertainty of the assigned value of CRM, reproducibility of the measurements made on CRM, inhomogeneity of CRM). The knowledge and skill of the analyst derived from experience and all of the currently available information are required for selecting the proper CRM and for evaluating the uncertainty of the measurement result.

References

1. Guide to the Expression of Uncertainty in Measurement, ISO, Geneva, 1993.
2. ISO Guide 30: Terms and Definitions Used in Connection with Reference Materials, ISO, Geneva, 1992.
3. ISO Guide 31: Contents of Certificates and Labels, ISO, Geneva, (in press).
4. ISO Guide 32: Calibration in Analytical Chemistry and Use of Certified Reference Materials, ISO, Geneva, 1997.
5. ISO Guide 33: Uses of Certified Reference Materials, ISO, Geneva, (in press)
6. ISO Guide 35: Certification of Reference Materials - General and Statistical Principles, ISO, Geneva, 1985 (under revision).
7. ISO 11095 :1994, Linear Calibration Using Reference Material, ISO, Geneva,
8. International Vocabulary of Basic and General Terms in Metrology, ISO, Geneva, 1993.
9. ISO 3534-1: 1993, Statistics - Vocabulary and Symbols - Part 1 : Probability and General Statistical Terms, ISO, Geneva.

10. ISO 5725 : 1994, Accuracy (Trueness and Precision) of Measurement Methods and Results, ISO, Geneva.

Overview on the Activities at ISO/REMCO

H. Klich

FEDERAL INSTITUTE FOR MATERIALS RESEARCH AND TESTING (BAM), RUDOWER CHAUSSEE 5, D-12489 BERLIN, GERMANY

REMCO is ISO's committee on **reference materials**, responsible to the ISO Technical Management Board.[1]

In addition to its basic activities within ISO, the main aim of the committee, since its inception in 1975 at the recommandation of an international seminar on certified reference materials (CRM), is to continue harmonizing CRMs and promoting their use worldwide.

Chairman: Harry Klich (1999)
Federal Institute for Materials
Research and Testing (BAM)
Rudower Chaussee 5
D-12489 Berlin, Germany

Phone: +49 30 6392 5847
Fax: +49 30 677706 10
Email: harry.klich@bam.de

Secretary: Jean-Remy Alessi
ISO Central Secretariat
1, rue de Varembé
CH-1211 Genève 20

Phone: +41 22 749 01 11
Fax: +41 22 733 34 30
Email: Alessi@isocs.iso.ch

OBJECTIVES

To establish definitions,
categories, levels and classifications of reference materials for use by ISO.

To determine the structure
of related forms of reference materials.

To formulate criteria
for choosing sources for mention in ISO documents (including legal aspects).

To prepare guidelines
for technical committees for making reference to reference materials in ISO documents.

To propose,
as far as necessary, action to be taken on reference materials required for ISO work.

To deal with matters
within the competence of the Committee, in relation with other international organizations and to advise the Technical Management Board on action to be taken.

WORK PROGRAMME

Hierarchy Task Group   Convenor: Prof P De Bièvre

Terms of reference

- To consider definitions, categories, levels and classification of reference materials, and recommend actions for REMCO deliberation.
- To provide models for the establishment of traceability of (certified) values in reference materials.
- To contribute to the revision process of VIM, the International vocabulary of basic and general terms in metrology.
- To revise ISO Guide 30:1992, Terms and definitions used in connection with reference materials as a result of the revision of VIM.

Calibration Task Group   Convenor: Dr. Hidetaka Imai

Terms of reference

- To study mathematical, including statistical, models of calibration using Certified Reference Materials (CRMs), and to draft appropriate guidance to CRM users for inclusion in ISO guides.
- To develop mathematical models that will assist CRM producers set certified levels when certification data arise from multiple methods, multiple laboratories, methoddependent analyses, or combinations of these sources.
- To collect documentation on calibration of instruments and methods by RMs.
- To revise ISO Guide 33, Uses of certified reference materials, and ISO Guide 35, Certification of reference materials-General and statistical principles.

Promotion Task Group   Convenor: Thomas E. Gills

Terms of reference

- To collect and disseminate databases of available CRMs and in-production (including planned) CRMs.
- To provide liaison with ISO and IEC technical committees (TCs), international organizations, institutions, agencies and CRM users to identify their needs for CRMs and convey them to producers.

- To inform TCs of CRM availability and encourage mention of CRMs in standards, as appropriate.
- To help organize workshops, seminars and demonstrations, partly in cooperation with DEVCO, to train potential users.
- To study future needs in connection with CRMs and prepare relevant propositions to the ISO Technical Managment Board.
- To assist development and dissemination of the index COde of Reference Materials (hence COMAR) to advance the promotion of reference materials.

Accreditation Task Group Convenor: Dr. Ronald Walker

Terms of reference

- To assess the need for the accreditation of RM producers.
- To collect, assess and analyse viewpoints and documentation concerning with the accreditation of RM producers.
- To provide liaison with appropriate national and international organizations concerned with the accreditation of RM producers.
- To coordinate future revisions of ISO Guide 34, Quality system guidelines for the production of reference materials.
- To draw up rules for accreditation of RM producers

Sampling Task Group   Convenor: Dr. Adriaan M.H. van der Veen

Terms of reference

- To collect relevant information with respect to the role of sampling and sub-sampling in the production and use of CRMs.
- To prepare draft annexes to relevant ISO Guides.
- To prepare a bibliography with relevant literature addressing matters on (sub-) sampling of materials.

Transportation and Distribution of Reference Materials
Convenor: William P. Reed

Terms of reference

- To develop an inventory of difficulties associated with the transportation and distribution of RMs.
- To identify and prioritize problem areas.

Members of REMCO:

P-members (participating)	O-members (observers)
Brazil (ABNT)	Argentina (IRAM)
Canada (SCC)	Australia (SAA)
China (CSBT)	Barbados (BNSI)
Czech Republic (COSMT)	Belgium (IBN)
Ecuador (INEN)	Brunei Darussalam (CPRU)
France (AFNOR)	Colombia (ICO NTEC)
Germany (DIN)	Croatia (DZNM)
Hungary (MSZT)	Cuba (NC)
India (BIS)	Denmark (DS)
Indonesia (DSN)	Egypt (EOS)
Iran, Islamic	Ethiopia (ESA)
Rep. of (ISIRI)	Finland (SFS)
Italy (UNI)	Greece (ELOT)
Japan (ISC)	Ireland (NSAI)
Korea, Rep. of (KNIT Q)	Israel (SII)
Netherlands (NNI)	Kenya (KEBS)
Poland (PKN)	Moldova (MOLDST)
Russian Federation (GOST R)	Mongolia (MNCSM)
Slovakia (UNMS)	New Zealand (SNZ)
South Africa (SABS)	Norway (NSF)
Sweden (SIS)	Portugal (IPQ)
Switzerland (SNV)	Romania (IRS)
United Kingdom (BSI)	Saudi Arabia (SASO)
USA (ANSI)	Tanzania (TBS)
	Thailand (TISI)
	Tunisia (INNORPI)
	Turkey (TSE)
	Ukraine (DSTU)
	Venezuela (COVENIN)
	Viet Nam (TCVN)
	Yugoslavia (SZS)

Organisations in liasion with REMCO

EC/IRMM/MT - European Commission - Institute for Reference Materials and Measurements/Measuring and Testing
COWS of WASP - Commission on World Standards of the World Association of Societies of Pathology
ECCLS - European Committee for Clinical Laboratory Standards
IAEA - International Atomic Energy Agency
IEC - International Electrotechnical Commission
IFCC - International Federation of Clinical Chemistry

ILAC - International Laboratory Accreditation Conference
IUPAC - International Union of Pure and Applied Chemistry
OIML - International Organization of Legal Metrology
UNEP-HEM - United Nations Environmental Programme, Harmonization of
Environmental Measurement
WHO - World Health Organization

ISO Guides developed by REMCO

ISO Guide 30: 1992, *Terms and definitions used in connection with reference materials.*[2]
It was drawn up by REMCO in the series of Guides in relation with guidelines for the
preparation, certification and use of reference materials (RMs) and certified reference
materials (CRMs). The first edition of this Guide (1981) was the outcome of collaboration
between REMCO and the organizations IEC, IAEA, OIML, IUPAC, IFCC and WHO. The
revision leading to the second edition was undertaken because it had become apparent that
some confusion existed as to what types *of measurement standards or etalons* should
legitimately be included within the definition of a reference material. Moreover, the
recognition that certified reference materials are measurement standards made it desirable
to examine the vocabulary of standards in metrology, as detailed in the *International
vocabulary of basic and general terms in metrology (VIM)*, Second edition (1993) with
particular reference to certified reference materials.

ISO Guide 31: 1981, *Contents of certificates of reference materials.* (Under revision, the
new version is expected in 1999)[3]
There are three types of different information (in the order of length) in connection with
certified reference materials (CRMs):
- the label on the container,
- the certificate and
- the report.

Normally, because the size of the label is limited by the container to which it is to be
affixed, the label can submit only the briefest information on the reference material.
Reference materials should be clearly labeled so that they are unambiguously identified
and referenced against accompanying certificates or other documentation.

The report on the reference material contains all the relevant information including all the details of the certification procedure.

The certificate, which can be described as the synopsis of the report, has normally one or two pages and should communicate information about a reference material from the producer to the user. This information is a statement of the certified property values, their meaning, and their uncertainties.

ISO Guide 32: 1997, *Calibration in analytical chemistry and use of certified reference materials*[4]

The calibration of the parameters associated with chemical analyses and material testing deserves particular attention, because major errors can be made by neglecting or ignoring the basic principles of metrology which also apply to this areas. This guide gives a number of general recommendations.

ISO Guide 33: 1989, *Uses of certified reference materials* (Under revision, the new version is expected in 1999)[5]

This guide explains the "proper use" and the "misuse" of certified reference materials (CRMs). The preparation of a CRM is a time-consuming and expensive endeavour. For this reason, CRMs must be used properly, i.e. effectively, efficiently and economically. There are many measurement processes where CRMs are in general use but are replaceable by a host of working standards such as homogeneous materials, previously analyzed materials, pure compounds solutions of pure elements, etc. The advantages in using CRMs are that the user has the means to assess the trueness and precision of his measurement method and establishes metrological traceability for his results. Whether the use of CRMs in these procedures is in fact "misuse" depends largely on the availability and relative cost of the CRMs. Where CRMs are in short supply or very expensive, their use would indeed be misuse. However, for CRMs in ample supply or where similar CRMs are available from one or more producers, it is strongly recommended that CRMs always be used instead of working standards because of the resultant enhanced confidence in the measurement output. The misuse of CRMs can occur also when the user does not fully take into account the uncertainty in the certified property. The overall uncertainty of a certified property of a CRM can have contributions from the inhomogeneity of the material, the within-laboratory uncertainty and, where applicable, the between-laboratories uncertainty. The user should not apply a CRM of greater uncertainty than permitted by the end-use. Guide 33 states that the uncertainty of the replicated measurement of a CRM is twice the between- laboratories standard deviation of the certification program, when this parameter is known, or four times the repeatability of the method on the CRM. These values represent the lowest level of uncertainty achievable with this CRM and may be used as a guide in determining whether it satisfies the uncertainty requirement of the end-use.

ISO Guide 34: 1996, Quality system guidelines for the production of reference materials.[6] This Guide is for the use of reference material producers in the development and implementation of their quality system, and by accreditation bodies, certifying bodies and others concerned with assessing the quality systems of reference material producers. A reference material needs to be characterized mainly to the level of accuracy required for its intended purpose (i.e. appropriate uncertainty). The reference material producer should describe the procedure for establishing the quality of the materials as a component of the quality system. The guide sets out the quality system requirements in accordance with which reference materials should be produced. It is intended to be used as part of a reference material producer' s general QA procedures.

ISO Guide 35: 1989, *Certification of reference materials -General and statistical principles.*[7]

The purpose of this Guide is to give a basic introduction to concepts and practical aspects related to the certification of reference materials. The quality of a measurement based on the use of a CRM will depend in part on the effort and care expended by the certification body on determining the property value(s) of the candidate CRM. Hence the process of certification should be carried out using well-characterized measurement methods that have high accuracy as well as precision and provide property values traceable to fundamental units of measurement.

Furthermore, the methods should yield values with uncertainties that are appropriate to the expected end-use of the CRM. Two clauses are devoted to the two most important technical considerations in the certification of CRMs - measurement uncertainties and material homogeneity. It assists in understanding valid methods for the certification of

RMs and also to help potential users to better define their technical requirements. The Guide should be useful in establishing the full potential of CRMs as aids to assuring the accuracy and interlaboratory compatibility of measurements on a national or international scale. In April 1998 REMCO forwarded the task to revise this Guide to its groups Calibration and Sampling.

Other activities:

The role of reference materials in achieving quality in analytical chemistry (1997)
This booklet deals with the application of reference materials and is written for those involved in the daily practice of analytical chemistry. It is free of charge and available on request from the ISO secretariat.

Introduction to the ISO/REMCO guides on reference materials (under development)
This document gives an brief overview on the existing ISO guides 30-35. The potenial user can decide which guide may be useful for his specifc purpose.

Worldwide Listing of CRM Producers
Since April 1997 accessable on the Internet URL: http://www.bam.de/a_i/comar/src/titel.htm
This database has been set up with the adresses of the producers stored already in the database for CRMs COMAR. Producers can be indentified by application fields and country.

References

1. Information on REMCO (1998)

2. **ISO Guide 30: 1992,** *Terms and definitions used in connection with reference materials.*

3. **ISO Guide 31: 1981,** *Contents of certificates of reference materials.*

4. **ISO Guide 32: 1997,** *Calibration in analytical chemistry and use of certified reference materials.*

5. **ISO Guide 33: 1989,** *Uses of certified reference materials*

6. **ISO Guide 34: 1996,** *Quality system guidelines for the production of reference materials.*

7. **ISO Guide 35: 1989,** *Certification of reference materials -General and statistical principles.*

Addresses of Authors

R. Bojanowski
Institute of Oceanology, PAS
81-712 Sopot
POLAND

K.I. Burns
International Atomic Energy Agency
Agency's Laboratories Seibersdorf
A-2444 Seibersdorf
AUSTRIA

P.R. Danesi
International Atomic Energy Agency
Agency's Laboratories Seibersdorf
A-2444 Seibersdorf
AUSTRIA

P. De Bievre
IRMM- Institute for Reference
Materials and Measurements
Retiesweg
B-2440 Geel
BELGIUM

P.P. De Regge
International Atomic Energy Agency
Agency's Laboratories Seibersdorf
A-2444 Seibersdorf
AUSTRIA

E. Deak
National Office of Measures (OMH)
P.O.Box 919
H-1535 Budapest
HUNGARY

A. Fajgelj
International Atomic Energy Agency
Agency's Laboratories Seibersdorf
A-2444 Seibersdorf
AUSTRIA

K. Fröhlich
International Atomic Energy Agency
Wagramer Strasse 5
A-1400 Vienna
AUSTRIA

T. Gills
National Institute of Standards and
Technology (NIST)
MD-20899-0001 Gaithersburg
USA

M. Gröning
International Atomic Energy Agency
Wagramer Strasse 5
A-1400 Vienna
AUSTRIA

M. Hedrich
Federal Institute of Materials Research and
Testing - BAM
Rudower Chaussee 5
D-12200 Berlin
GERMANY

H.-J. Heine
Federal Institute of Materials
Research and Testing - BAM
Rudower Chaussee 5
D-12200 Berlin
GERMANY

H. Klich
Federal Institute for Materials Research and
Testing - BAM
Rudower Chaussee 5
D-12489 Berlin
GERMANY

J. LaRosa
International Atomic Energy Agency
Marine Environment Laboratory
4. Quai Antoine 1er
MC 98012 B.P. 800
MONACO

P. Mader
Faculty of Agronomy, Department of
Chemistry
Czech University of Agriculture
Suchdol
CZ-165 21 Prague
CZECH REPUBLIC

J. Moreno Bermudez
International Atomic Energy Agency
Agency's Laboratories Seibersdorf
A-2444 Seibersdorf
AUSTRIA

K. Okamoto
National Institute of Materials and
Chemical Research
1-1 Higashi, Tsakuba
305-8565 Ibaraki
JAPAN

Pan Xiu Rong
Chinese National Research Centre for
Certified Reference Materials
No. 18, Bei San Huan Dong Lu,
Chaoyangqu
100013 Beijing
CHINA

J. Pauwels
Institute for Reference Materials and
Measurements (IRMM)
Management of Reference Materials Unit
Retieseweg
B-2440 Geel
BELGIUM

Z. Radecki
International Atomic Energy Agency
Agency's Laboratories Seibersdorf
A-2444 Seibersdorf
AUSTRIA

H.F. Steger
Canada Centre for Mineral and Energy
Technology
Energy, Mine & Resources Canada
555 Booth Street
ON K1A 0G1 Ottawa
CANADA

M. Suchanek
Institute of Chemical Technology
Department of Analytical Chemistry
166 28 Prague
CZECH REPUBLIC

A.M.H. van der Veen
Department of Chemistry
P.O.Box 654
NL-2600 AR Delft
THE NETHERLANDS

R.F. Walker
Laboratory of the Government
Chemist - LGC
Queens Road
Teddington TW11 0LY
UNITED KINGDOM

T. Win
Federal Institute of Materials Research and
Testing - BAM
Rudower Chaussee 5
D-12489 Berlin
GERMANY

J. Yoshinaga
National Institute for environmental
Studies
16-2, Onogawa, Tsukuba
305-0053 Ibaraki
JAPAN

Zhao Min
Chinese National Research Centre for
Certified Reference Materials
No. 18, Bei San Huan Dong Lu,
Chaoyangqu
100013 Beijing
CHINA

Subject Index

Acceptance criterion, 184, 185
Accreditation program, 1, 2, 6
Accuracy, 5, 8, 13, 18, 23, 25, 29, 46, 58, 59, 71, 73, 75, 83, 104, 115, 120, 129, 130-139, 141, 147, 151, 152, 167, 194, 195
level of, 120
Ambient atmosphere measurements, 100
Amount of substance, 72, 77, 173, 177
Analyte, 16, 59, 66, 69, 73, 76, 143, 144, 151, 163-165, 167
AQCS (Analytical Quality Control Services), *See also* IAEA, 65, 80, 81

BERM (International Symposium on Biological and Environmental Reference Materials), 32, 41, 93
Bias, *See also* systematic error, 8, 9, 16, 29, 58, 59, 104, 105, 139, 148, 151-158, 182, 183, 185, 186
Bureau Communautaire de Référence (BCR), 31- 45, 47, 51, 52, 80, 96

Calibrant (Calibration standard), 33
Calibration curve, 24, 28, 72, 98, 139, 141, 172, 175
Calibration gases, *See also* Reference material (RM), gaseous, 115, 117, 119, 121-123, 125
Canadian Certified Reference Materials Project (CCRMP), 127, 128, 142
Certification
campaign, 71, 93, 94, 183
plan, 9

Certification exercise, *See also* Certification procedure, 34
Certification report, 33, 35, 37, 40, 130, 142
Certified reference material (CRM), 30, 31, 41, 44, 46, 57, 67, 72, 100-102, 107, 108, 130, 131, 133, 141-143, 151, 163, 165, 176, 177, 179, 180, 188, 189, 192, 193, 196
certificate, 1, 2, 5, 10, 24, 25, 35, 41, 67, 124, 130, 144, 145, 163, 172, 177, 180-182, 186, 192, 193
for chemical composition, 31
for coal analysis, 100
for isotope ratio measurements, 31
for quality control, 100
for reactor neutron dosimetry, 31
for trace elements, 32
isotopic, 31
matrix, 3, 16, 18-21, 24-26, 28, 29, 32, 33, 35, 36, 46, 47, 57, 59, 60, 65, 66, 71-73, 76, 79, 81, 93, 95, 96, 100, 101, 105, 107, 117, 118, 127-133, 137, 139, 140-142, 151-154, 156, 161, 163, 170, 171, 175, 181, 186
natural biological, 32
spiked, 31
Certifying agency, 32
Chain of comparisons, 2, 71, 130, 173
Chemical composition, 18, 82
Chemical engineering, 1
Chemicals, 31, 127, 143
high purity, 31
China Law on Metrology, 1